# Instructor's Manual with Answer Key and Complete Testing Program for

# Geometric Tolerancing
## A Text-Workbook

**Second Edition**

### Richard S. Marrelli
Late Professor of Industrial Technology
Los Angeles Pierce College
Woodlawn Hills, California

### Patrick J. M<sup>c</sup>Cuistion, Ph.D.
Department of Industrial Technology
Ohio University
Athens, Ohio

**GLENCOE**
McGraw-Hill

New York, New York    Columbus, Ohio    Mission Hills, California    Peoria, Illinois

# CONTENTS

**Introduction** 1

**Daily Schedules** 2

    Semester course, 3–4; Quarter course, 5

**Periodic Tests** 7

|  | Test | Chapters |  | Test Page | Answer Page |
|---|---|---|---|---|---|
| SEMESTER | 1 | 1–6 | Introduction, Symbols, Terms, Datums, Inspection of Geometric Tolerances, General Rules | 9 | 43 |
|  | 2 | 7–10 | Straightness, Flatness, Circularity, Cylindricity | 13 | 45 |
|  | 3 | 11–14 | Profile, Parallelism, Perpendicularity, Angularity | 15 | 46 |
|  | 4 | 15, 16 | Concentricity and Runout | 17 | 47 |
|  | 5 | 17–20 | Position Tolerance—General and Location Applications Position-Coaxial Applications and Symmetry | 19 | 48 |
| QUARTER | 1Q | 1–6 | Introduction, Symbols, Terms, Datums, Inspection of Geometric Tolerances, General Rules | 23 | 49 |
|  | 2Q | 7–11 | Straightness, Flatness, Circularity, Cylindricity, Profile | 25 | 50 |
|  | 3Q | 12–16 | Parallelism, Perpendicularity, Angularity, Concentricity, and Runout | 27 | 51 |
|  | 4Q | 17, 18 | Position Tolerance—General and Location Applications | 29 | 51 |

**Final Examination** 31

**Answers to Periodic Tests** 41

**Answers to Final Examination** 53

**Answers to Chapter Review Problems** 59

    **1** (61)   **5** (65)   **9** (70)   **13** (75)   **17** (80)
    **2** (61)   **6** (65)   **10** (71)  **14** (76)  **18** (82)
    **3** (62)   **7** (67)   **11** (72)  **15** (78)  **19** (85)
    **4** (63)   **8** (68)   **12** (74)  **16** (79)  **20** (87)

**Answers to Comprehensive Exercises** 89

    **1** (91), **2** (92), **3** (93), **4** (93), **5** (94), **6** (94)

*Instructor's Manual with Answer Key and Complete Testing Program for*
***Geometric Tolerancing: A Text–Workbook, Second Edition***

Copyright © 1997 by Glencoe/McGraw-Hill. All rights reserved. Except as permitted under the United States Copyright Act, no part of this publication may be reproduced or distributed in any form or by any means, or stored in a database or retrieval system, without the prior written permission of the publisher.

Send all inquiries to:
Glencoe/McGraw-Hill
936 Eastwind Dr.
Westerville, OH 43081        2 3 4 5 6 7 8 9 10 11 MAL 99 98 97 96

ISBN 0-02-801889-3      Printed in the United States of America.

# Introduction

Using the text-workbook *Geometric Tolerancing* and the material in this manual, an instructor teaching a semester or quarter program can initiate a course in geometric tolerancing with a minimum of preparation time. Only a knowledge of the subject and familiarization with the text are required. The standard schedules can easily be modified to suit longer or shorter courses and seminars. The material in this manual includes

1. Daily schedules for a semester course and a quarter course, pages 2–5.
2. Tear-out copy masters for five periodic tests for a semester course, pages 9–21.
3. Tear-out copy masters for four periodic tests for a quarter course, pages 23–29.
4. A tear-out copy master for a final examination to be used for both courses, pages 31–39.
5. Answer keys for the periodic tests (pages 41–51) and final examination (pages 53–58).
6. Answer keys for the workbook problems, pages 59–94.

# Daily Schedules

Schedules for each meeting of a semester course and a quarter course are on pages 3–5 of this manual. These can be modified by the instructor for longer or shorter courses.

The semester course covers all of the material in the text-workbook, including all the chapter review problems (RP) and comprehensive exercises (CE). Copy masters are provided for five periodic tests and a final examination.

The quarter course does *not* treat the entire text. The following topics have been deleted:

| | |
|---|---|
| Section 9.4 | Measuring Circularity Error |
| Section 18.8 | Use of Position Tolerance for Angularity at MMC |
| Section 18.10 | Multiple Patterns of Features |
| Section 18.11 | Composite Tolerances for Feature Patterns |
| Section 20.6 | Step Datums |
| Section 20.7 | Equalizing Datums |
| Appendix A | Summary of Nonstandard Practices |

The lesser course content does not fully compensate for the fewer hours, making the quarter course a more intensive course, with more lecture each day and more homework. There are four periodic tests, identified as 1Q, 2Q, 3Q, and 4Q, all shorter than the semester-course tests. The instructor may wish to lighten the workload by assigning only Part I or Part II of the chapter review problems.

Daily Schedule—Semester Course

| WK | Date | Collect Review Problems | Return & Review | Lecture 1 | Lecture 2 | Reading Assignment | Problem Assignment |
|---|---|---|---|---|---|---|---|
| 1 | | | | Overview of course<br>CH 1 Introduction | CH 2 Symbols<br>CH 3 Terms | Preface, CH 1, 2, 3 | RP 1<br>RP 2<br>RP 3-I |
| 2 | | RP 1<br>RP 2<br>RP 3-I | | CH 3 Terms<br>CH 4 Datums | CH 4 Datums<br>CH 5 Inspection | CH 3, 4, 5 | RP 3-II<br>RP 4-I & II<br>RP 5 |
| 3 | | RP 3-II<br>RP 4-I & II<br>RP 5 | RP 1<br>RP 2<br>RP 3-I | CH 6 General rules | CH 7 Straightness | CH 6, 7 | RP 6-I & II<br>RP 7-I & II |
| 4 | | RP 6-I & II<br>RP 7-I & II | RP 3-II<br>RP 4-I & II<br>RP 5 | TEST 1   CH 1–6 | CH 8 Flatness | CH 8 | RP 8-I & II |
| 5 | | RP 8-I & II | RP 6-I & II<br>RP 7-I & II | CH 9 Circularity | CH 10 Cylindricity | CH 9, 10 | RP 9-I & II<br>RP 10-I & II |
| 6 | | RP 9-I & II<br>RP 10-I & II | RP 8-I & II | TEST 2   CH 7–10 | CH 11 Profile | CH 11 | RP 11-I |
| 7 | | RP 11-I | RP 9-I & II<br>RP 10-I & II | CH 11 Profile | CH 12 Parallelism | CH 11, 12 | RP 11-II<br>RP 12-I & II |
| 8 | | RP 11-II<br>RP 12-I & II | RP 11-I | CH 13 Perpendicularity | CH 14 Angularity | CH 13, 14 | RP 13-I & II<br>RP 14-I & II |
| 9 | | RP 13-I & II<br>RP 14-I & II | RP 11-II<br>RP 12-I & II | TEST 3   CH 11–14 | CH 15 Concentricity | CH 15 | RP 15-I & II |
| 10 | | RP 15-I & II | RP 13-I & II<br>RP 14-I & II | CH 16 Runout | CH 16 Runout | CH 16 | RP 16-I & II |

*(over)*

© Glencoe/McGraw-Hill

Daily Schedule—Semester Course

| WK | Date | Collect Review Problems | Return & Review | Lecture 1 | Lecture 2 | Reading Assignment | Problem Assignment |
|---|---|---|---|---|---|---|---|
| 11 | | RP 16-I & II | RP 15-I & II | TEST 4  CH 15–16 | CH 17 Position—General 17.1–17.5 | CH 17 (17.1–17.5) | RP 17-I |
| 12 | | RP 17-I | RP 16-I & II | CH 17 Position—General 17.6–17.7 | CH 18 Position—Location 18.1–18.4 | CH 17 (17.6–17.7) CH 18 (18.1–18.4) | RP 17-II |
| 13 | | RP 17-II | RP 17-I | CH 18 Position—Location 18.5–18.9 | CH 18 Position—Location 18.10–18.11 | CH 18 (18.5–18.11) | RP 18-I & II |
| 14 | | RP 18-I & II | RP 17-II | CH 19 Position—Coaxial | CH 19 Position—Coaxial | CH 19 | RP 19-I |
| 15 | | RP 19-I | RP 18-I & II | CH 20 Symmetry | TEST 5  CH 17–20 | CH 20 | RP 19-II RP 20 |
| 16 | | RP 19-II RP 20 | RP 19-I | FINAL EXAM | | | |

**4**   Semester Course Schedule

© Glencoe/McGraw-Hill

## Daily Schedule—Quarter Course

| WK | Date | Collect Review Problems | Return & Review | Lecture 1 | Lecture 2 | Lecture 3 | Reading Assignment | Problem Assignment |
|---|---|---|---|---|---|---|---|---|
| 1 | | | | Overview of course CH 1 Introduction | CH 2 Symbols CH 3 Terms | CH 3 Terms CH 4 Datums | Preface CH 1, 2, 3, 4 | RP 1, RP 2 RP 3-I & II, RP 4-I |
| 2 | | RP 1, RP 2 RP 3-I & II, RP 4-I | | CH 4 Datums CH 5 Inspection | CH 6 General rules | TEST 1Q  CH 1–6 | CH 4, 5, 6 | RP 4-II RP 5 RP 6-I & II |
| 3 | | RP 4-II RP 5 RP 6-I & II | RP 1, RP 2 RP 3-I & II, RP 4-I | CH 7 Straightness | CH 8 Flatness | CH 9 Circularity | CH 7, 8, 9 | RP 7-I & II RP 8-I & II RP 9-I & II |
| 4 | | RP 7-I & II RP 8-I & II RP 9-I & II | RP 4-II RP 5 RP 6-I & II | CH 10 Cylindricity | CH 11 Profile | CH 11 Profile | CH 10, 11 | RP 10-I & II RP 11-I & II |
| 5 | | RP 10-I & II RP 11-I & II | RP 7-I & II RP 8-I & II RP 9-I & II | TEST 2Q  CH 7–11 | CH 12 Parallelism | CH 13 Perpendicularity | CH 12, 13 | RP 12-I & II RP 13-I & II |
| 6 | | RP 12-I & II RP 13-I & II | RP 10-I & II RP 11-I & II | CH 14 Angularity | CH 15 Concentricity | CH 16 Runout | CH 14, 15, 16 | RP 14-I & II RP 15-I & II RP 16-I |
| 7 | | RP 14-I & II RP 15-I & II RP 16-I | RP 12-I & II RP 13-I & II | CH 16 Runout | TEST 3Q  CH 12–16 | CH 17 Position— General 17.1–17.5 | CH 16, 17 | RP 16-II RP 17-I |
| 8 | | RP 16-II RP 17-I | RP 14-I & II RP 15-I & II RP 16-I | CH 17 Position— General 17.6–17.7 | CH 18 Position— Location 18.1–18.4 | CH 18 Position— Location 18.5–18.9 | CH 17, 18 | RP 17-II RP 18-I |
| 9 | | RP 17-II RP 18-I | RP 16-II RP 17-I | CH 18 Position— Location 18.10–18.11 | TEST 4Q  CH 17–18 | CH 19 Position— Coaxial | CH 18, 19 | RP 18-II RP 19-I |
| 10 | | RP 18-II RP 19-I | RP 17-II RP 18-I | CH 19 Position— Coaxial | CH 20 Symmetry | | CH 19, 20 | RP 19-II RP 20 |
| 11 | | RP 19-II RP 20 | RP 18-II RP 19-I & II RP 20 | FINAL EXAM | | | | |

# Periodic Tests

|  | Test | Chapters |  | Test Page | Answer Page |
|---|---|---|---|---|---|
| SEMESTER | 1 | 1–6 | Introduction, Symbols, Terms, Datums, Inspection of Geometric Tolerances, General Rules | 9 | 43 |
| | 2 | 7–10 | Straightness, Flatness, Circularity, Cylindricity | 13 | 45 |
| | 3 | 11–14 | Profile, Parallelism, Perpendicularity, Angularity | 15 | 46 |
| | 4 | 15, 16 | Concentricity and Runout | 17 | 47 |
| | 5 | 17–20 | Position Tolerance—General and Location Applications Position-Coaxial Applications and Symmetry | 19 | 48 |
| QUARTER | 1Q | 1–6 | Introduction, Symbols, Terms, Datums, Inspection of Geometric Tolerances, General Rules | 23 | 49 |
| | 2Q | 7–11 | Straightness, Flatness, Circularity, Cylindricity, Profile | 25 | 50 |
| | 3Q | 12–16 | Parallelism, Perpendicularity, Angularity, Concentricity, and Runout | 27 | 51 |
| | 4Q | 17, 18 | Position Tolerance—General and Location Applications | 29 | 51 |

Name _____ Date _____

# Test 1: Chapters 1–6
## Introduction, Symbols, Terms, Datums, Inspection of Geometric Tolerances, General Rules

This is a short-answer recall test in which you will write a statement or complete a sentence in the blank space. When the required answer is one word, the blank is short. For two words there are two short blanks. When a statement is required, the blank is about as long as the required answer; however, the answer does not have to be that length to be correct.

1. What does the abbreviation ANSI stand for?
   _____

2. How was geometric tolerance specified on drawings before the introduction of the symbol system? _____

3. List six geometric characteristics and draw the correct symbol next to each.

   _____ ____      _____ ____

   _____ ____      _____ ____

   _____ ____      _____ ____

4. Draw full size an example of a datum feature symbol (also called a datum identifying symbol).




5. Three letters are not used in datum feature symbols. Name *two* of them.

   _____   _____

6. There are four modifier symbols. Draw the symbols for *three* of them.

   _____   _____   _____

7. A feature control frame may contain five items of information. Name *three* of them.
   _____

8. Make a sketch of a simple bar and add a feature control frame for straightness with a tolerance of .002. Omit dimensions.

*(over)*

9. How is a feature control frame applied to a surface when it is not practical to attach it to an extension line? _____

10. Which item in the feature control frame determines whether a manufacturing error is acceptable? _____

11. A theoretically exact feature from which dimensions may be taken is called a(n) _____.

12. How does a datum *plane* differ from a datum *feature*? _____

13. If more than one datum is used, the first one is the most important in the _____ of the part.

14. What is meant by "mating parts"? _____

15. In a hole, what is the condition called when the hole is produced at the lower limit? (The whole term or an abbreviation may be given.) _____

16. The size of an object that results in the tightest fit with the mating part is called the _____ condition.

17. FIM is the abbreviation for _____.

18. On drawings that do not specify geometric tolerances, what determines the geometric form? _____

19. The pitch diameter rule applies to gears, splines, and _____.

20. Datum targets are used mostly with parts having rough, uneven surfaces, such as _____ and _____.

21. An unsupported object in space can be moved in six directions, technically denoted as _____.

22. In a datum frame, how many datum target points are required for the second datum? _____

23. The designer selects datum planes on a particular part based on the _____ of the part and its _____ in the next assembly.

10    Periodic Test 1

Name _____ Date _____

24. What is the diameter of a datum target symbol on a drawing using .125 tall lettering?

   _____

25. Where is the letter of the datum plane and the number of the tooling point given on the drawing?

   _____

26. How is a datum target *point* represented on the drawing?

   _____

27. When a datum target *area* is circular, where is its diameter specified?

   _____

28. A cylindrical surface is equivalent in locating a part to how many datum planes?

   _____

29. A datum plane that consists of two parallel surfaces separated by an offset is called

   a(n) _____ datum.

30. When a datum target is on the far side of a view (not visible), how is this shown on the drawing?

   _____

© Glencoe/McGraw-Hill

Name _____  Date _____

# Test 2: Chapters 7–10
## Straightness, Flatness, Circularity, Cylindricity

This is a short-answer recall test in which you will write a statement or complete a sentence in the blank space. When the required answer is one word, the blank is short. For two words there are two short blanks. When a statement is required, the blank is about as long as the required answer; however, the answer does not have to be that length to be correct.

1. Straightness error is the measure of how much each straight surface element or the axis of an object deviates in one direction from being a(n) _____

   _____.

2. How does flatness differ from straightness?

   _____

3. Circularity error is the measure of how much each circular element of an object deviates from being a(n) _____.

4. How does cylindricity differ from circularity?

   _____

5. Draw the correct symbol for each geometric characteristic below.

   Circularity _____    Straightness _____

   Cylindricity _____    Flatness _____

6. Which modifier applies to straightness, unless otherwise specified. _____

7. Element straightness is specified by attaching the feature control frame to an extension line. How is the feature control frame applied when *size* straightness is required?

   _____

8. Draw an example of a feature control frame for a unit straightness of .001 per inch of length.

(over)

© Glencoe/McGraw-Hill                    Periodic Test 2    13

9. To the requirement of Problem 8, add a total straightness tolerance of .010.

10. You are designing a heavy casting and wish to specify flatness only in a .50 × 1.00 area of a large surface. Make a sketch below of the controlled area and specify a flatness tolerance of .001. Omit dimensions.

11. The tolerance zone for circularity is the _____ distance between two concentric circles.

12. In the space below, make a single-view sketch of a cylinder .50 diameter by .75 long, and apply a circularity tolerance of .005. Omit dimensions.

13. A shaft has a diameter of 1.1250–1.1253, and no circularity tolerance is specified.

    What is the maximum permissible circularity error? _____

14. Circularity error is measured in a plane perpendicular to the _____ of the circular feature.

15. Cylindricity is a combination of which two geometric characteristics?

    _____

16. A cylindrical tolerance zone is equivalent to a(n) _____ tolerance zone curled into the shape of a cylinder.

17. Straightness, flatness, circularity, and cylindricity tolerance are all required to be

    within dimensional _____ _____. (Rule 1)

18. Testing cylindrical error requires measuring two kinds of elements on the surface:

    _____ and _____ elements.

Name _____ Date _____

# Test 3: Chapters 11–14
## Profile, Parallelism, Perpendicularity, Angularity

This is a short-answer recall test in which you will write a statement or complete a sentence in the blank space. When the required answer is one word, the blank is short. For two words there are two short blanks. When a statement is required, the blank is about as long as the required answer; however, the answer does not have to be that length to be correct.

1. Profile of a line tolerance controls just the _____ of a curved surface in one direction, not *all* of the _____. (Same word.)

2. Profile of a surface tolerance differs from profile of a line tolerance in that profile of a surface tolerance controls _____.

3. When are two lines parallel? _____
_____

4. The tolerance zone for profile of a line tolerance is the space between imaginary curved lines. The tolerance zone for profile of a surface tolerance is the space between two imaginary curved _____.

5. In profile tolerances, the tolerance is assumed to be bilateral, unless _____.

6. Draw an example of a feature control frame specifying a profile of a surface tolerance of .002 relative to a datum, between two points designated A and B on the drawing.

7. An automotive valve cam has a profile of a surface tolerance of .05 mm that applies all around. No datum is necessary. Draw an appropriate feature control frame.

8. When a profile tolerance is used to control coplanarity of two flat surfaces, the leader points to _____.

9. Parallelism may be applied to other geometric forms besides lines. Name *two* of them. _____  _____

*(over)*

© Glencoe/McGraw-Hill    Periodic Test 3    15

10. When a parallelism tolerance zone is an imaginary cylinder, the tolerance is specified as a diameter. With which type of feature does this occur? _____

11. A plate that is 5.0–5.1 mm thick has a parallelism tolerance of .05 mm. What is the maximum permissible out-of-flatness of the surfaces? _____

12. Perpendicularity is a special case of which other geometric characteristic?

   _____

13. A perpendicularity tolerance also controls _____.

14. Why is a tolerance expressed in degrees less desirable than an angularity tolerance?

   _____

15. The angle given in an angularity tolerance is basic (no tolerance), but some manufacturing error must be allowed. Where is the tolerance specified?

   _____.

16. Which other geometric characteristic is used with size features in preference to an angularity tolerance at MMC? _____

17. Since parallelism, perpendicularity, and angularity are relationship characteristics, a(n) _____ is always needed.

18. When a feature is a symmetrical size feature such as a hole, how is the feature control frame applied in the drawing?

   _____

19. Give an example of a size feature with which a parallelism, perpendicularity, or angularity tolerance may apply at MMC. _____

20. A feature control frame may be attached to an extension line or dimension line, but never to a(n) _____ _____.

Name _____  Date _____

# Test 4: Chapters 15 and 16
# Concentricity and Runout

This is a short-answer recall test in which you will write a statement or complete a sentence in the blank space. When the required answer is one word, the blank is short. For two words there are two short blanks. When a statement is required, the blank is about as long as the required answer; however, the answer does not have to be that length to be correct.

1. Two cylinders are coaxial when they have the same _____.

2. There is no symbol for the general case of coaxiality. Draw the symbols that are used for three special cases.

   _____    _____    _____

3. Concentricity error is the amount by which the axes of two regular solids, such as cylinders, are _____.

4. Give one reason why concentricity is not as often specified as runout and position tolerance. _____

5. The tolerance zone for concentricity is always an imaginary cylinder. How is this fact shown in the feature control frame?
   _____

6. If the concentricity of a cylinder is ⌀.03 mm, what is the maximum slant, in millimeters, of the feature axis relative to the datum axis? _____

7. Concentricity does not include errors in surface characteristics. Name *one* of these characteristics. _____

8. Give an example of a machine part in which concentricity should be controlled.
   _____

9. Runout is any deviation of a surface from perfect form that can be detected by _____.

10. Runout is a composite tolerance, including errors in other geometric characteristics. Name *one*. _____

(over)

© Glencoe/McGraw-Hill                                   Periodic Test 4    17

11. In circular runout, the deviation of each _____ _____ is controlled.

12. Total runout controls the deviation of all _____ in a surface, circular and straight.

13. Why should circular runout be specified, rather than total runout, whenever it is adequate?
    _____

14. Two separated cylinders on a crankshaft are to be used as one datum for a total runout tolerance of .02 mm. Draw the required feature control frame.

15. When is it necessary to specify a flat face as a datum in addition to a cylindrical surface? _____
    _____

16. Why is it necessary to place a stop against the opposite face of the part when inspecting a flat face for runout?
    _____

17. A(n) _____ is required with concentricity, runout, and position tolerance because these are relationship features.

18. What material modifier applies to concentricity and runout, unless otherwise specified?
    _____

19. A high-speed turbine rotor shaft must maintain very close coaxiality in order to minimize imbalance. The designer decides this is an RFS application, but which coaxiality tolerance should be specified? Why?
    _____
    _____

Name _____  Date _____

# Test 5: Chapters 17–20
# Position Tolerance—General and Location Applications, Position-Coaxial Applications and Symmetry

This is a short-answer recall test in which you will write a statement or complete a sentence in the blank space. When the required answer is one word, the blank is short. For two words there are two short blanks. When a statement is required, the blank is about as long as the required answer; however, the answer does not have to be that length to be correct.

All questions have a grading point value of one unless otherwise specified. Total points = 30.

1. Position *error* is any _____ in the location of a feature.

   Position *tolerance* is the total _____ in the location of a feature. (2 points)

2. Like runout, position tolerance is a composite tolerance, including surface errors in addition to location error. Name *two* of the possible surface errors.

   _____, _____

3. For a feature located by its axis, such as a round hole, the shape of the position tolerance zone is an imaginary _____.

4. For a feature located by its center plane, such as a rectangular slot, the position tolerance zone is the space between two imaginary _____.

5. Coordinate dimensioning results in a square or rectangular tolerance zone in which the permissible error is not the same in all directions. This is corrected in position tolerance, since the tolerance zone is _____ in shape.

6. A basic dimension is an exact value with no specified _____.

7. The _____ for a basic dimension is given in the _____ _____.

8. What type of tolerance is often controlled when using position, in addition to location?

   _____

9. The LMC modifier is most often used to control a critical _____ _____.

*(over)*

© Glencoe/McGraw-Hill        Periodic Test 5    19

**10.** How does an RFS modifier affect the specified position tolerance?

_____

**11.** In a floating fastener assembly, what is allowed to float relative to what?

_____

**12.** A projected tolerance zone transfers the position tolerance from the part being controlled to the _____ _____.

**13.** Where the fasteners are studs or press-fit pins, the projected tolerance zone must be equal not to the thickness of one of the parts but to the _____
_____.

**14.** When a position tolerance specification is converted to zero tolerance at MMC, the position tolerance is reduced to zero, but this is compensated for by increasing the
_____.

**15.** In designing the fit of a tab in a slot, the total tolerance in both parts is made equal to the minimum clearance between the mating features. Write the formula used for the calculation. _____

**16.** Position tolerance is used in place of angularity for a size feature when what modifier is applied? _____

**17.** Symmetry error is the amount by which opposite sides of a size feature are unequally spaced from the _____ _____ of the datum.

**18.** Two or more feature patterns on the same part are called
_____.

**19.** Multiple hole patterns located by basic dimensions from the same datums are considered one _____ _____.

**20.** How many gages are used to inspect a composite hole pattern? _____

**21.** A plate has three different hole patterns, and each one is to be gaged separately. What note is added below the feature control frame? _____

**22.** The form of the tolerance zone for coaxial position tolerance is an imaginary

_____.

Name _____ Date _____

**23.** In situations where maintaining a minimum wall thickness is more critical than coaxiality, which modifier is specified? _____

**24.** An engine block has five coaxial holes to support a camshaft. The common axis must be positioned relative to three surfaces of the block within .005 diameter, while the axes of the individual holes must be all within .0005 diameter, completely within the .005 diameter. Draw a composite feature control frame to specify this design intent.

**25.** In order to avoid interference in fitting coaxial parts, the total position tolerance for both parts must not be greater than the _____ _____ between the parts.

**26.** When a coaxiality tolerance for a bore is specified as zero at MMC, where does the machinist find the actual manufacturing tolerance?

_____

**27.** Two parts having mating coaxial features fit together. Which type of coaxial control should the designer specify on the two detail drawings? _____

Name _____   Date _____

# Test 1Q: Chapters 1–6
# Introduction, Symbols, Terms, Datums,
# Inspection of Geometric Tolerances, General Rules

This is a short-answer recall test in which you will write a statement or complete a sentence in the blank space. When the required answer is one word, the blank is short. For two words there are two short blanks. When a statement is required, the blank is about as long as the required answer; however, the answer does not have to be that length to be correct.

1. How was geometric tolerance specified on drawings before the introduction of the symbol system? _____

2. List six geometric characteristics and draw the correct symbol next to each.

   _____ _____      _____ _____

   _____ _____      _____ _____

   _____ _____      _____ _____

3. Draw full size an example of a datum feature symbol (also called a datum identifying symbol).

4. There are four modifier symbols. Draw the symbols for *three* of them.

   _____   _____   _____

5. Make a sketch of a simple bar and add a feature control frame for straightness with a tolerance of .002. Omit dimensions.

6. An element may be a straight line or it may be a(n) _____.

7. How does a datum *plane* differ from a datum *feature*? _____

   _____

8. The size of an object that results in the tightest fit with the mating part is called the _____ condition.

9. On drawings that do not specify geometric tolerances, what determines the geometric form? _____

© Glencoe/McGraw-Hill         Periodic Test 1Q   23

Name _____  Date _____

# Test 2Q: Chapters 7–11
# Straightness, Flatness,
# Circularity, Cylindricity, Profile

This is a short-answer recall test in which you will write a statement or complete a sentence in the blank space. When the required answer is one word, the blank is short. For two words there are two short blanks. When a statement is required, the blank is about as long as the required answer; however, the answer does not have to be that length to be correct.

1. How does flatness differ from straightness? _____
   _____

2. Circularity error is the measure of how much each circular element of an object deviates from being a(n) _____.

3. How does cylindricity differ from circularity? _____
   _____

4. Draw the correct symbol for each geometric characteristic. Circularity _____

   Straightness _____  Flatness _____  Cylindricity _____

5. Which modifier applies to straightness unless otherwise specified? _____

6. Draw an example of a feature control frame for a unit straightness of .001 per inch of length.

7. You are designing a heavy casting and wish to specify flatness only in a .50 × 1.00 area of a large surface. Make a sketch below of the controlled area and specify a flatness tolerance of .001. Omit dimensions.

8. Cylindricity is a combination of which two geometric characteristics?
   _____

*(over)*

© Glencoe/McGraw-Hill                                      Periodic Test 2Q    25

9. In profile tolerances, the tolerance is assumed to be bilateral, unless _____.

10. An automotive valve cam has a profile of a surface tolerance of .05 mm that applies all around. No datum is necessary. Draw an appropriate feature control frame.

Name _____  Date _____

# Test 3Q: Chapters 12–16
# Parallelism, Perpendicularity, Angularity, Concentricity, and Runout

This is a short-answer recall test in which you will write a statement or complete a sentence in the blank space. When the required answer is one word, the blank is short. For two words there are two short blanks. When a statement is required, the blank is about as long as the required answer; however, the answer does not have to be that length to be correct.

1. When are two lines parallel? _____
   _____.

2. When a parallelism tolerance zone is an imaginary cylinder, the tolerance is specified as a diameter. With which type of feature does this occur?

   _____

3. Perpendicularity is a special case of which other geometric characteristic?

   _____

4. The angle given in an angularity tolerance is basic (no tolerance), but some manufacturing error must be allowed. Where is the tolerance specified?

   _____

5. Concentricity tolerance is the maximum permissible error in

   _____.

6. If the concentricity of a cylinder is ⌀.03 mm, what is the maximum slant, in millimeters, of the feature axis relative to the datum axis? _____

7. Runout is any deviation of a surface from perfect form that can be detected by

   _____.

8. Total runout is a composite tolerance, including errors in other geometric characteristics. Name *one*. _____

Name _____   Date _____

# Test 4Q: Chapters 17 and 18
# Position Tolerance—General and Location Applications

This is a short-answer recall test in which you will write a statement or complete a sentence in the blank space. When the required answer is one word, the blank is short. For two words there are two short blanks. When a statement is required, the blank is about as long as the required answer; however, the answer does not have to be that length to be correct.

1. Position *error* is any _____ in the location of a feature.

   Position *tolerance* is the total _____ _____ in the location of a feature. (2 points)

2. Coordinate dimensioning results in a square or rectangular tolerance zone in which the permissible error is not the same in all directions. This is corrected in position tolerance, since the tolerance zone is _____ in shape.

3. In a floating fastener assembly, what is allowed to float relative to what?

   _____

4. A basic dimension is an exact value with no specified _____.

5. When a position tolerance specification is converted to zero tolerance at MMC, the position tolerance is reduced to zero, but this is compensated for by increasing the

   _____.

6. In designing the fit of a tab in a slot, the total tolerance in both parts is made equal to the minimum clearance between the mating features. Write the formula used for the calculation. _____

7. In addition to location, what type of tolerance is often controlled when using position? _____

© Glencoe/McGraw-Hill                    Periodic Test 4Q    29

Name _____  Date _____

**39.** (10 points.)
A batch of cover plates, shown below, was inspected, and the holes were found to be well within the position tolerance but undersize by .25 mm; therefore not acceptable. However, the inspector, who understood the concept of zero position tolerance at MMC, suggested a change in the feature control frame on the drawing, which would make the parts acceptable without rework. Draw in the space below the revised feature control frame and hole callout suggested by the inspector.

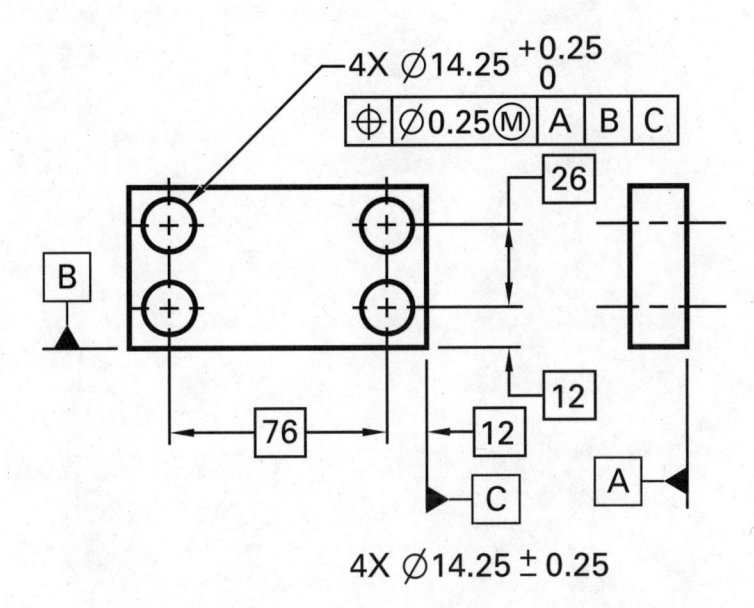

4X ⌀14.25 $^{+0.25}_{0}$

| ⊕ | ⌀0.25 Ⓜ | A | B | C |

4X ⌀14.25 ± 0.25

# Answers to Periodic Tests

|  | Test | Chapters |  | Student Test Page | Answer Page |
|---|---|---|---|---|---|
| **SEMESTER** | 1 | 1–6 | Introduction, Symbols, Terms, Datums, Inspection of Geometric Tolerances, General Rules | 9 | 43 |
|  | 2 | 7–10 | Straightness, Flatness, Circularity, Cylindricity | 13 | 45 |
|  | 3 | 11–14 | Profile, Parallelism, Perpendicularity, Angularity | 15 | 46 |
|  | 4 | 15, 16 | Concentricity and Runout | 17 | 47 |
|  | 5 | 17–20 | Position Tolerance—General and Location Applications Position-Coaxial Applications and Symmetry | 19 | 48 |
| **QUARTER** | 1Q | 1–6 | Introduction, Symbols, Terms, Datums, Inspection of Geometric Tolerances, General Rules | 23 | 49 |
|  | 2Q | 7–11 | Straightness, Flatness, Circularity, Cylindricity, Profile | 25 | 50 |
|  | 3Q | 12–16 | Parallelism, Perpendicularity, Angularity, Concentricity, and Runout | 27 | 51 |
|  | 4Q | 17, 18 | Position Tolerance—General and Location Applications | 29 | 51 |

# Test 1: Chapters 1–6
## Introduction, Symbols, Terms, Datums, Inspection of Geometric Tolerances, General Rules

This is a short-answer recall test in which the student is expected to write a statement or complete a sentence in the blank space. When the required answer is one word, the blank is short. For two words there are two short blanks. When a statement is required, the blank is about as long as the required answer; however, the answer does not have to be that length to be correct.

1. What does the abbreviation ANSI stand for?
   *American National Standards Institute*

2. How was geometric tolerance specified on drawings before the introduction of the symbol system? *By Local Notes*

3. List six geometric characteristics and draw the correct symbol next to each.
   *(Any six in text Fig. 2–1, pp. 5–7)*

4. Draw full size an example of a datum feature symbol (also called a datum identifying symbol).
   [A]◄

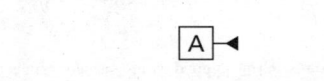

5. Three letters are not used in datum feature symbols. Name *two* of them.
   *I    O    Q*

6. There are four modifier symbols. Draw the symbols for *three* of them.
   Ⓜ Ⓛ Ⓟ

7. A feature control frame may contain five items of information. Name *three* of them.
   *Geometric characteristic; tolerance; tolerance modifier; datum; datum modifier*

8. Make a sketch of a simple bar and add a feature control frame for straightness with a tolerance of .002. Omit dimensions.
   [ — | .002 ]

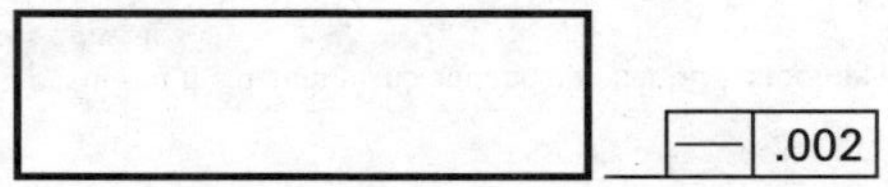

9. How is a feature control frame applied to a surface when it is not practical to attach it to an extension line? *By a leader from the frame to the surface*

10. Which item in the feature control frame determines whether a manufacturing error is acceptable? *The tolerance*

11. A theoretically exact feature from which dimensions may be taken is called a(n) *Datum*

12. How does a datum *plane* differ from a datum *feature*? *The datum plane is perfect; the datum feature is an actual imperfect surface*

13. If more than one datum is used, the first one is the most important in the *Fit (Function)* of the part.

14. What is meant by "mating parts"? *Parts that fit together*

15. In a hole, what is the condition called when the hole is produced at the lower limit? *MMC* (The whole term or an abbreviation may be given.)

16. The size of an object that results in the tightest fit with the mating part is called the *virtual* condition.

17. FIM is the abbreviation for *Full Indicator Movement*

18. On drawings that do not specify geometric tolerances, what determines the geometric form? *The size tolerances*

19. The pitch diameter rule applies to gears, splines, and *screw threads*

20. Datum targets are used mostly with parts having rough, uneven surfaces, such as *castings* and *forgings*.

21. An unsupported object in space can be moved in six directions, technically denoted as *six degrees of freedom*

22. In a datum frame, how many datum target points are required for the second datum? *Two*

23. The designer selects datum planes on a particular part based on the *function* of the part and its *fit* in the next assembly.

24. What is the diameter of a datum target symbol on a drawing using .125 tall lettering?

    *.44*

25. Where is the letter of the datum plane and the number of the tooling point given on the drawing?

    *In the lower half of the datum target symbol*

26. How is a datum target *point* represented on the drawing?

    *By a 45° cross (X)*

27. When a datum target *area* is circular, where is its diameter specified?

    *In the upper half of the datum target symbol*

28. A cylindrical surface is equivalent in locating a part to how many datum planes?

    *Two*

29. A datum plane that consists of two parallel surfaces separated by an offset is called a(n) ___*step*___ datum.

30. When a datum target is on the far side of a view (not visible), how is this shown on the drawing?

    *The leader is a hidden line*

# Test 2: Chapters 7–10
## Straightness, Flatness, Circularity, Cylindricity

This is a short-answer recall test in which the student is expected to write a statement or complete a sentence in the blank space. When the required answer is one word, the blank is short. For two words there are two short blanks. When a statement is required, the blank is about as long as the required answer; however, the answer does not have to be that length to be correct.

1. Straightness error is the measure of how much each straight surface element or the axis of an object deviates in one direction from being a(n) _____*straight*_____ _____*line*_____.

2. How does flatness differ from straightness? _____*Flatness controls all the elements in all directions*_____

3. Circularity error is the measure of how much each circular element of an object deviates from being a(n) _____*circle*_____.

4. How does cylindricity differ from circularity? _____*Cylindricity controls straight elements as well as circular elements*_____

5. Draw the correct symbol for each geometric characteristic below.

   Circularity ○  Straightness —
   Cylindricity ⌭  Flatness ▱

6. Which modifier applies to straightness, unless otherwise specified. _____*RFS*_____

7. Element straightness is specified by attaching the feature control frame to an extension line. How is the feature control frame applied when *size* straightness is required? _____*To the size dimension or its dimension line*_____

8. Draw an example of a feature control frame for a unit straightness of .001 per inch of length.

   | — | .001/1.000 |

9. To the requirement of Problem 8, add a total straightness tolerance of .010.

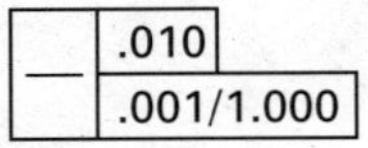

   | — | .010 |
   |   | .001/1.000 |

10. You are designing a heavy casting and wish to specify flatness only in a .50 × 1.00 area of a large surface. Make a sketch below of the controlled area and specify a flatness tolerance of .001. Omit dimensions.

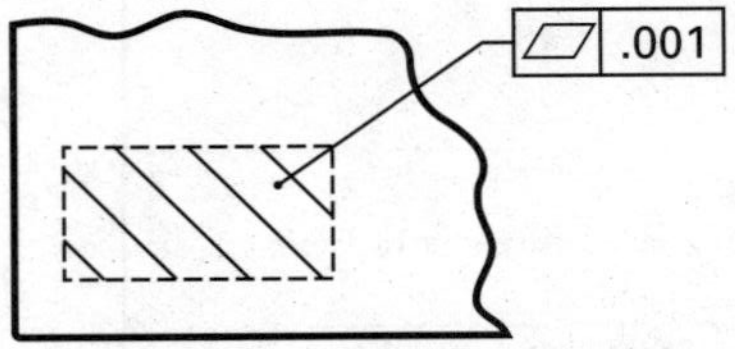

    [sketch showing hatched rectangular area with ▱ .001 callout]

11. The tolerance zone for circularity is the _____*radial*_____ distance between two concentric circles.

12. In the space below, make a single-view sketch of a cylinder .50 diameter by .75 long, and apply a circularity tolerance of .005. Omit dimensions.

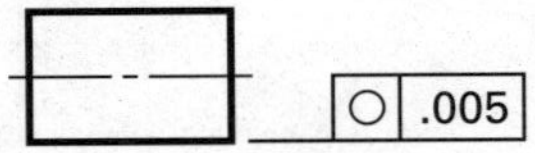

    | ○ | .005 |

13. A shaft has a diameter of 1.1250–1.1253, and no circularity tolerance is specified. What is the maximum permissible circularity error? _____*.0003*_____

14. Circularity error is measured in a plane perpendicular to the _____*axis*_____ of the circular feature.

15. Cylindricity is a combination of which two geometric characteristics? _____*Circularity and straightness*_____

16. A cylindrical tolerance zone is equivalent to a(n) _____*flatness*_____ tolerance zone curled into the shape of a cylinder.

17. Straightness, flatness, circularity, and cylindricity tolerance are all required to be within dimensional _____*size*_____ _____*tolerances (limits)*_____ (Rule 1).

18. Testing cylindrical error requires measuring two kinds of elements on the surface: _____*circular*_____ and _____*straight*_____ elements.

# Test 3: Chapters 11–14
## Profile, Parallelism, Perpendicularity, Angularity

This is a short-answer recall test in which the student is expected to write a statement or complete a sentence in the blank space. When the required answer is one word, the blank is short. When a statement is required, the blank is about as long as the required answer; however, the answer does not have to be that length to be correct.

1. Profile of a line tolerance controls just the _____elements_____ of a curved surface in one direction, not *all* of the _____elements_____ (Same word.)

2. Profile of a surface tolerance differs from profile of a line tolerance in that profile of a surface tolerance controls _____the entire surface (all of the elements)_____

3. When are two lines parallel? _____When every point on one line is the same distance from opposing points on another line._____

4. The tolerance zone for profile of a line tolerance is the space between imaginary curved lines. The tolerance zone for profile of a surface tolerance is the space between two imaginary curved _____surfaces_____

5. In profile tolerances, the tolerance is assumed to be bilateral, unless _____otherwise specified_____

6. Draw an example of a feature control frame specifying a profile of a surface tolerance of .002 relative to a datum, between two points designated A and B on the drawing.

7. An automotive valve cam has a profile of a surface tolerance of .05 mm that applies all around. No datum is necessary. Draw an appropriate feature control frame.

8. When a profile tolerance is used to control coplanarity of two flat surfaces, the leader points to _____an extension line between the surfaces_____

9. Parallelism may be applied to other geometric forms besides lines. Name *two* of them. _____Planes_____  _____Cylinders_____  _____Cones_____

10. When a parallelism tolerance zone is an imaginary cylinder, the tolerance is specified as a diameter. With which type of feature does this occur? _____Cylinder_____

11. A plate that is 5.0–5.1 mm thick has a parallelism tolerance of .05 mm. What is the maximum permissible out-of-flatness of the surfaces? _____0.05_____

12. Perpendicularity is a special case of which other geometric characteristic? _____Angularity_____

13. A perpendicularity tolerance also controls _____flatness_____

14. Why is a tolerance expressed in degrees less desirable than an angularity tolerance? _____It is not uniform in width_____

15. The angle given in an angularity tolerance is basic (no tolerance), but some manufacturing error must be allowed. Where is the tolerance specified? _____In the feature control frame_____

16. Which other geometric characteristic is used with size features in preference to an angularity tolerance at MMC? _____Position tolerance_____

17. Since parallelism, perpendicularity, and angularity are relationship characteristics, a(n) _____datum_____ is always needed.

18. When a feature is a symmetrical size feature such as a hole, how is the feature control frame applied in the drawing? _____Below or to one side of the dimension or attached to the dimension line_____

19. Give an example of a size feature with which a parallelism, perpendicularity, or angularity tolerance may apply at MMC. _____a tab (or a slot or hole)_____

20. A feature control frame may be attached to an extension line or dimension line, but never to a(n) _____center_____ _____line_____

# Final Examination

Answers, pages 53–58.

Name _____ Date _____

# Final Examination
# Geometric Tolerancing

Write **T** in the space at the left if you believe the statement is true, or **F** if you believe it is false.

_____ 1. Geometric tolerance is the total permissible variation in the shape or location of a feature and/or in its relationship to other features.

_____ 2. The maximum material condition of a hole is always its largest permissible size.

_____ 3. When no geometric tolerance is specified, the dimensional tolerance controls the form of the part as well as its size.

_____ 4. Unless otherwise specified, all straightness tolerances apply at MMC.

_____ 5. Cylindricity is a combination of circularity and straightness.

_____ 6. Unless otherwise specified, profile tolerances are assumed to be bilateral.

_____ 7. Parallelism and perpendicularity tolerances do not require a datum.

_____ 8. Runout is any deviation of a surface from perfect form that can be detected by taking diametral measurements at 90°.

_____ 9. In a floating fastener assembly, both parts have clearance holes.

In problems 10–29, write the letter of the best answer in the space at left.

_____ 10. If the height of a block has a size tolerance of ±.010 and the bottom is absolutely flat, the upper surface can be curved by a maximum of _____.

(a) .005   (b) .010   (c) .020   (d) more than .020

_____ 11. The virtual condition of an object is the size that results in the _____ fit with a mating part.

(a) loosest   (b) average   (c) tightest   (d) nominal

_____ 12. A good feature to select for a datum would be a(n) _____.

(a) lathe center   (b) bearing surface   (c) center line   (d) axis

_____ 13. In making a dial indicator test for parallelism, the FIM is .004. This indicates a parallelism error of _____.

(a) .002   (b) .004   (c) .008   (d) more than .008

_____ 14. In addition to screw threads, the pitch diameter rule also applies to splines and _____.

(a) gears   (b) pulleys   (c) knurling   (d) bearings

(over)

© Glencoe/McGraw-Hill                                                            Final Examination   33

15. There are two varieties of straightness: element straightness and _____ straightness.
    (a) size feature  (b) cylindrical  (c) square  (d) conical

16. A flatness tolerance zone is the space between two parallel _____.
    (a) planes  (b) elements  (c) straight lines  (d) datums

17. The shape of the tolerance zone for circularity is most like a(n) _____.
    (a) cylinder  (b) radius  (c) ellipse  (d) ring or washer

18. The shape of the tolerance zone for cylindricity is most like a _____.
    (a) doughnut  (b) globe  (c) tube  (d) cylinder

19. Profile of a line tolerance controls not an entire surface but only _____.
    (a) chords  (b) radii  (c) elements  (d) arcs

20. Any tolerance zone that is cylindrical in shape is designated in the feature control frame by a symbol meaning _____.
    (a) parallel  (b) cylindricity  (c) diameter  (d) area

21. If no modifier is specified in the feature control frame, perpendicularity applies _____.
    (a) at LMC  (b) at MMC  (c) at virtual condition  (d) RFS

22. When an angularity tolerance is specified, the angle dimension must be _____.
    (a) basic  (b) unilateral  (c) bilateral  (d) unidirectional

23. The word _____ is used to describe the general case of the geometric characteristic where axes are in line.
    (a) concentricity  (b) alignment  (c) coaxiality  (d) runout

24. Runout is a _____ tolerance, including errors in circularity, straightness, perpendicularity, and coaxiality.
    (a) composite  (b) complex  (c) multiple  (d) position

25. Position tolerance is the total _____ error in the location of a feature relative to another feature or to several other features.
    (a) measurable  (b) permissible  (c) unilateral  (d) bilateral

26. In a fixed fastener assembly, one of the mating parts has a _____ hole; the other or others have clearance holes.
    (a) tight-fitting  (b) loose-fitting  (c) transition-fit  (d) lubricated

27. The tolerance for coaxial position is always expressed as a _____.
    (a) radius  (b) diameter  (c) half-width  (d) total width

Name _____ Date _____

_____ 28. The specified points where contact is made between the tooling and the part are called _____.
   (a) datum targets   (c) supporting points
   (b) datum features  (d) tooling targets

_____ 29. A system of three datum planes at right angles is called a _____.
   (a) tooling setup   (c) position fix
   (b) datum frame     (d) manufacturing frame

Most of the following problems require drawing to be done by the student. This may be mechanical drawing or freehand sketching.

**30.** (7 points.)
In the space below, draw one view of a rectangular part with two parallel holes, side by side. Omit dimensions. Show that the hole on the right, regardless of its size, is parallel with the other hole within .003. (Hint: What is the shape of the tolerance zone?)

**31.** (6 points.)
A .375–.380 diameter hole has a parallelism tolerance of .006 at MMC relative to another hole, RFS. In the space below, draw a feature control frame to specify this situation.

**32.** (5 points.)
A valve stem consists of two cylinders that must be positioned within .0005 diameter regardless of the produced size of both cylinders. Draw an appropriate feature control frame.

(over)

© Glencoe/McGraw-Hill          Final Examination    35

**33.** (3 points.)

A link has five 10 mm ±.05 mm punched holes in a line, with a position tolerance of ⌀.05 mm at MMC. Complete the permissible variation table below for the possible produced sizes shown.

| Produced Size of Hole | Permissible Position Error |
|---|---|
| ⌀10.05 mm | _____ |
| 10.00 | _____ |
| 9.95 | _____ |

**34.** (15 points.)

The drawing below shows a front view and a left side view of a cover plate. The entire surface of the curved profile must be as dimensioned from the bottom surface and perpendicular to the back surface within the following tolerances.

Between A and B: .005
B and C: .004
C and D: .002

Complete the drawing.

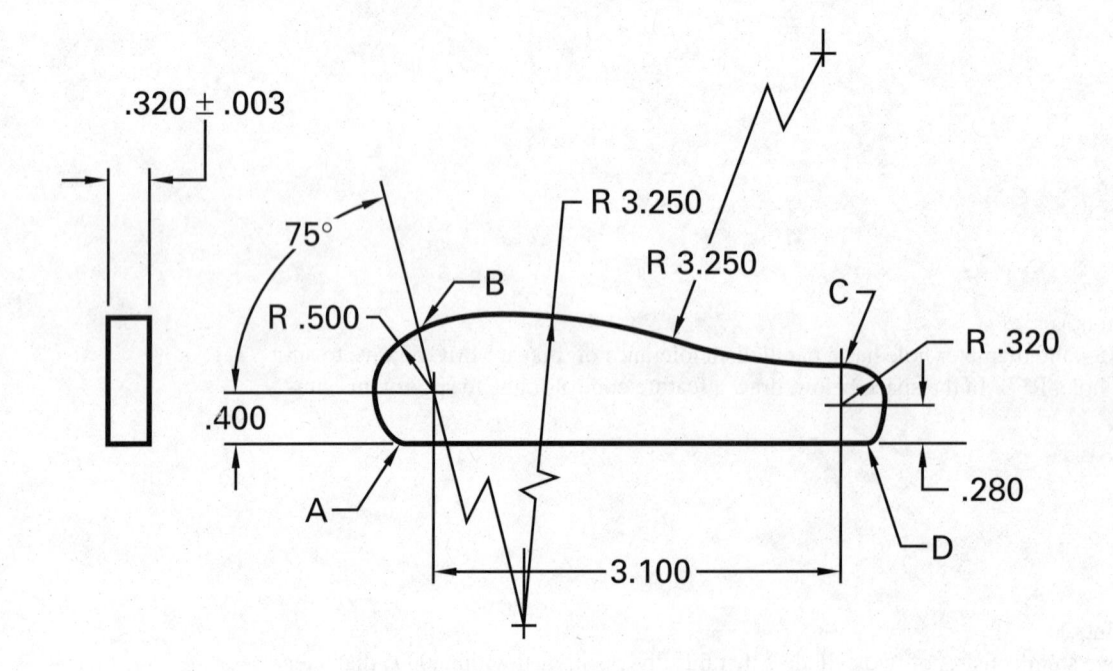

36  **Final Examination**  © *Glencoe/McGraw-Hill*

Name _____  Date _____

**35.** (13 points.)
The common axis of the four holes in the hinge body drawn below must be within a diametral tolerance zone of .015 at MMC relative to the front and top surfaces (location tolerance). At the same time all four holes must be coaxial within ⌀.001 at MMC. Complete the drawing.

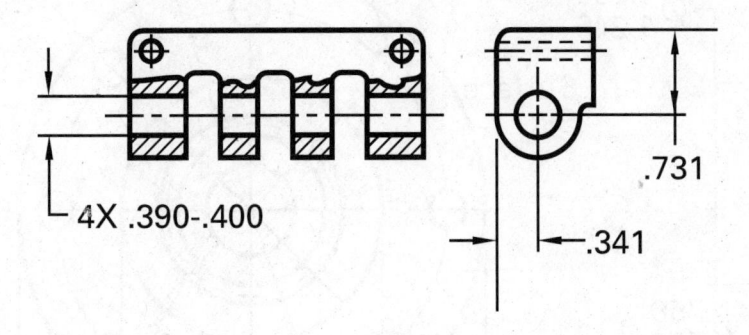

4X .390-.400

.731

.341

**36.** (46 points.)
In the two-view drawing of a threaded flange on the next page, add appropriate symbology to comply with the given geometric requirements. The one-digit numbers (1–6) are not part of the drawing; they are used only for identification purposes in this problem. Remember to combine datum feature symbols with feature control frames whenever possible.

(a) Surface 1 to be flat within .001

(b) Surface 2 to be parallel with Surface 1 within .0025

(c) Diameter 3 to be perpendicular to Surface 1 within ⌀.0015

(d) Circular runout of Diameter 4 relative to Diameter 3 to be within .004

(e) The screw thread to be coaxial with Diameter 3 within .004 when both diameters are at MMC

(f) The ⌀.312 holes to be positioned relative to Surface 1 and Diameter 5 within ⌀.005 when the diameters are at MMC

(g) The total runout of Diameter 5 relative to Diameter 4 to be within .003

(h) Diameter 6 to be round within .010

(over)

© Glencoe/McGraw-Hill                                                                                   Final Examination    37

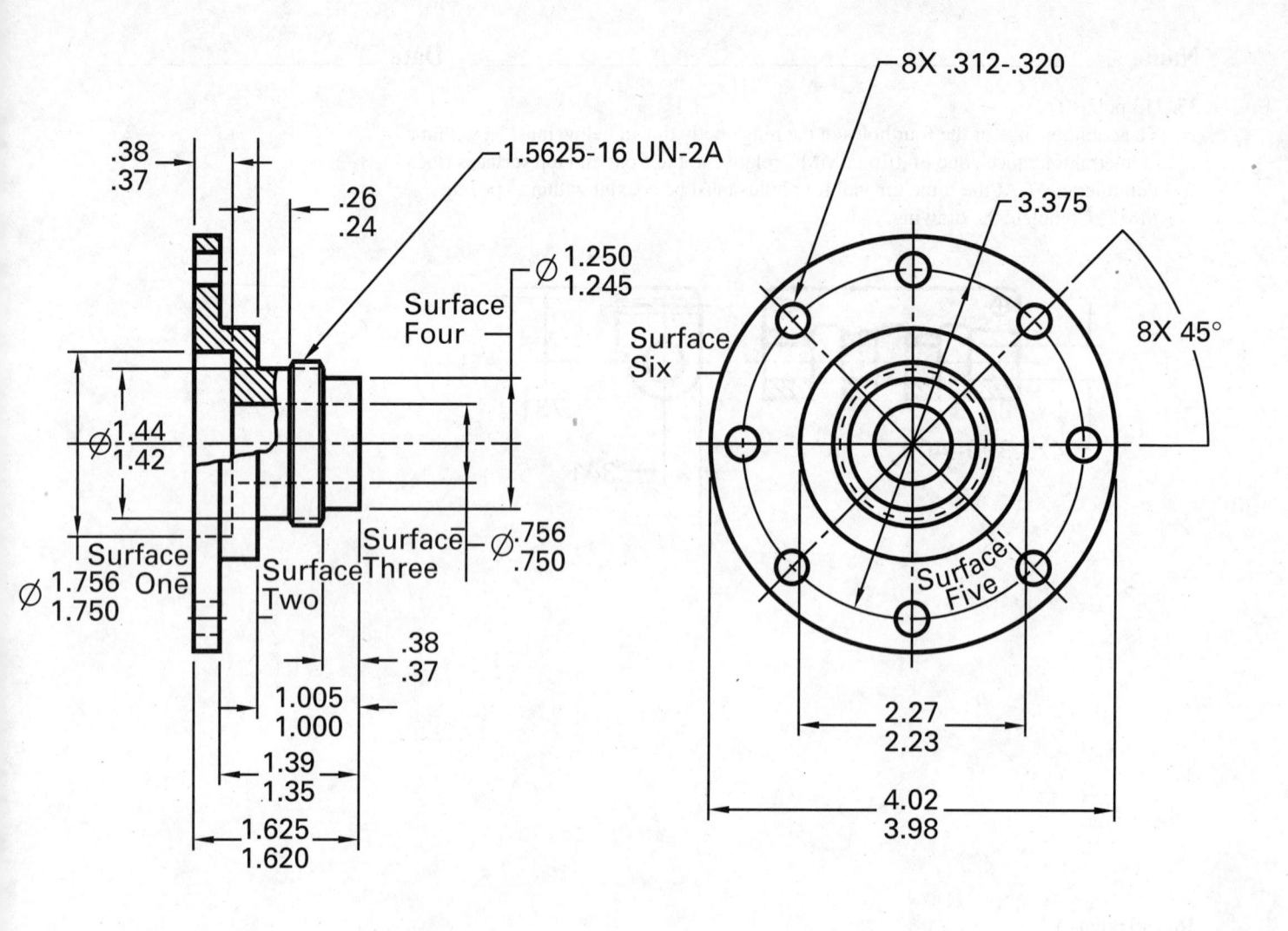

**37.** (5 points.)
What are the virtual conditions of clearance holes in the drawing of a cylinder head to be installed over ⌀.375 × 1.25-high studs in the mating part when the position tolerance for studs and holes is ⌀.020 at MMC, relative to the front and one side edge? Show all calculations. _____

**38.** (8 points.)
Draw the feature control frame for the clearance holes in Problem 38.

# Final Examination
## Geometric Tolerancing

Write **T** in the space at the left if you believe the statement is true, or **F** if you believe it is false.

_T_  1. Geometric tolerance is the total permissible variation in the shape or location of a feature and/or in its relationship to other features.

_F_  2. The maximum material condition of a hole is always its largest permissible size.

_T_  3. When no geometric tolerance is specified, the dimensional tolerance controls the form of the part as well as its size.

_F_  4. Unless otherwise specified, all straightness tolerances apply at MMC.

_T_  5. Cylindricity is a combination of circularity and straightness.

_T_  6. Unless otherwise specified, profile tolerances are assumed to be bilateral.

_F_  7. Parallelism and perpendicularity tolerances do not require a datum.

_F_  8. Runout is any deviation of a surface from perfect form that can be detected by taking diametral measurements at 90°.

_T_  9. In a floating fastener assembly, both parts have clearance holes.

In problems 10–29, write the letter of the best answer in the space at left.

_c_  10. If the height of a block has a size tolerance of ±.010 and the bottom is absolutely flat, the upper surface can be curved by a maximum of _____.
(a) .005  (b) .010  (c) .020  (d) more than .020

_c_  11. The virtual condition of an object is the size that results in the _____ fit with a mating part.
(a) loosest  (b) average  (c) tightest  (d) nominal

_b_  12. A good feature to select for a datum would be a(n) _____.
(a) lathe center  (b) bearing surface  (c) center line  (d) axis

_b_  13. In making a dial indicator test for parallelism, the FIM is .004. This indicates a parallelism error of _____.
(a) .002  (b) .004  (c) .008  (d) more than .008

_a_  14. In addition to screw threads, the pitch diameter rule also applies to splines and _____.
(a) gears  (b) pulleys  (c) knurling  (d) bearings

_a_  15. There are two varieties of straightness: element straightness and _____ straightness.
(a) size feature  (b) cylindrical  (c) square  (d) conical

_a_  16. A flatness tolerance zone is the space between two parallel _____.
(a) planes  (b) elements  (c) straight lines  (d) datums

_d_  17. The shape of the tolerance zone for circularity is most like a(n) _____.
(a) cylinder  (b) radius  (c) ellipse  (d) ring or washer

_c_  18. The shape of the tolerance zone for cylindricity is most like a _____.
(a) doughnut  (b) globe  (c) tube  (d) cylinder

_c_  19. Profile of a line tolerance controls not an entire surface but only _____.
(a) chords  (b) radii  (c) elements  (d) arcs

_c_  20. Any tolerance zone that is cylindrical in shape is designated in the feature control frame by a symbol meaning _____.
(a) parallel  (b) cylindricity  (c) diameter  (d) area

_d_  21. If no modifier is specified in the feature control frame, perpendicularity applies _____.
(a) at LMC  (b) at MMC  (c) at virtual condition  (d) RFS

_a_  22. When an angularity tolerance is specified, the angle dimension must be _____.
(a) basic  (b) unilateral  (c) bilateral  (d) unidirectional

_c_  23. The word _____ is used to describe the general case of the geometric characteristic where axes are in line.
(a) concentricity  (b) alignment  (c) coaxiality  (d) runout

_a_  24. Runout is a _____ tolerance, including errors in circularity, straightness, perpendicularity, and coaxiality.
(a) composite  (b) complex  (c) multiple  (d) position

_b_  25. Position tolerance is the total _____ error in the location of a feature relative to another feature or to several other features.
(a) measurable  (b) permissible  (c) unilateral  (d) bilateral

_a_  26. In a fixed fastener assembly, one of the mating parts has a _____ hole; the other or others have clearance holes.
(a) tight-fitting  (b) loose-fitting  (c) transition-fit  (d) lubricated

_b_  27. The tolerance for coaxial position is always expressed as a _____.
(a) radius  (b) diameter  (c) half-width  (d) total width

© Glencoe/McGraw-Hill

*a* _____ 28. The specified points where contact is made between the tooling and the part are called _____.
   (a) datum targets   (c) supporting points
   (b) datum features  (d) tooling targets

*b* _____ 29. A system of three datum planes at right angles is called a _____.
   (a) tooling setup   (c) position fix
   (b) datum frame     (d) manufacturing frame

Most of the following problems require drawing to be done by the student. This may be mechanical drawing or freehand sketching.

**30.** (7 points.)
In the space below, draw one view of a rectangular part with two parallel holes, side by side. Omit dimensions. Show that the hole on the right, regardless of its size, is parallel with the other hole within .003. (Hint: What is the shape of the tolerance zone?)

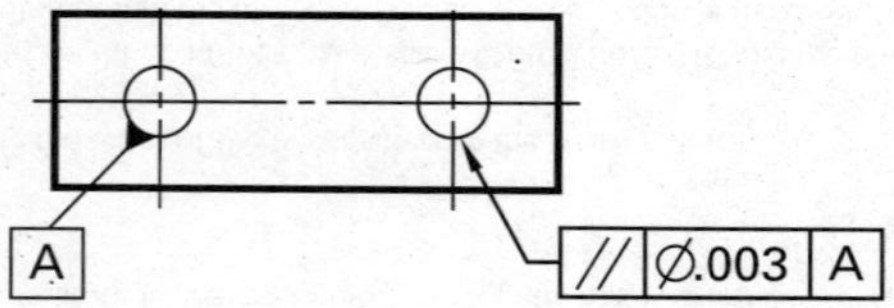

**31.** (6 points.)
A .375–.380 diameter hole has a parallelism tolerance of .006 at MMC relative to another hole, RFS. In the space below, draw a feature control frame to specify this situation.

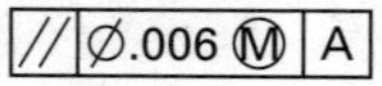

**32.** (5 points.)
A valve stem consists of two cylinders that must be positioned within .0005 diameter regardless of the produced size of both cylinders. Draw an appropriate feature control frame.

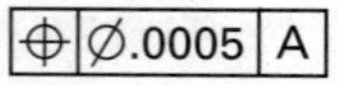

**33.** (3 points.)
A link has five 10 mm ±.05 mm punched holes in a line, with a position tolerance of Ø.05 mm at MMC. Complete the permissible variation table below for the possible produced sizes shown.

| Produced Size of Hole | Permissible Position Error |
|---|---|
| Ø10.05 mm | 0.15 |
| 10.00 | 0.10 |
| 9.95 | 0.05 |

**34.** (15 points.)
The drawing below shows a front view and a left side view of a cover plate. The entire surface of the curved profile must be as dimensioned from the bottom surface and perpendicular to the back surface within the following tolerances.
Between A and B: .005
         B and C: .004
         C and D: .002
Complete the drawing.

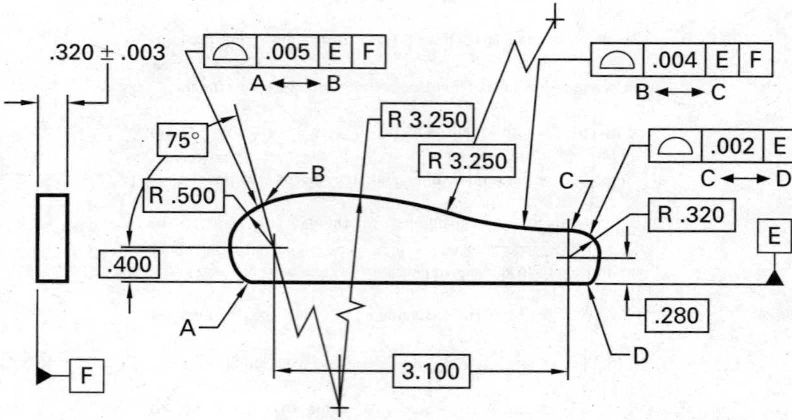

56   **Final Examination, Answers**                                    © Glencoe/McGraw-Hill

**35.** (13 points.)

The common axis of the four holes in the hinge body drawn below must be within a diametral tolerance zone of .015 at MMC relative to the front and top surfaces (location tolerance). At the same time all four holes must be coaxial within ⌀.001 at MMC. Complete the drawing.

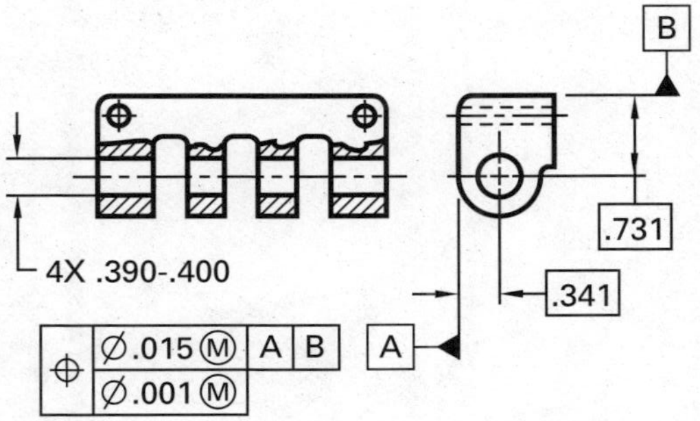

**36.** (46 points.)

In the two-view drawing of a threaded flange on the next page, add appropriate symbology to comply with the given geometric requirements. The one-digit numbers (1–6) are not part of the drawing; they are used only for identification purposes in this problem. Remember to combine datum feature symbols with feature control frames whenever possible.

(a) Surface 1 to be flat within .001
(b) Surface 2 to be parallel with Surface 1 within .0025
(c) Diameter 3 to be perpendicular to Surface 1 within ⌀.0015
(d) Circular runout of Diameter 4 relative to Diameter 3 to be within .004
(e) The screw thread to be coaxial with Diameter 3 within .004 when both diameters are at MMC
(f) The ⌀.312 holes to be positioned relative to Surface 1 and Diameter 5 within ⌀.005 when the diameters are at MMC
(g) The total runout of Diameter 5 relative to Diameter 4 to be within .003
(h) Diameter 6 to be round within .010

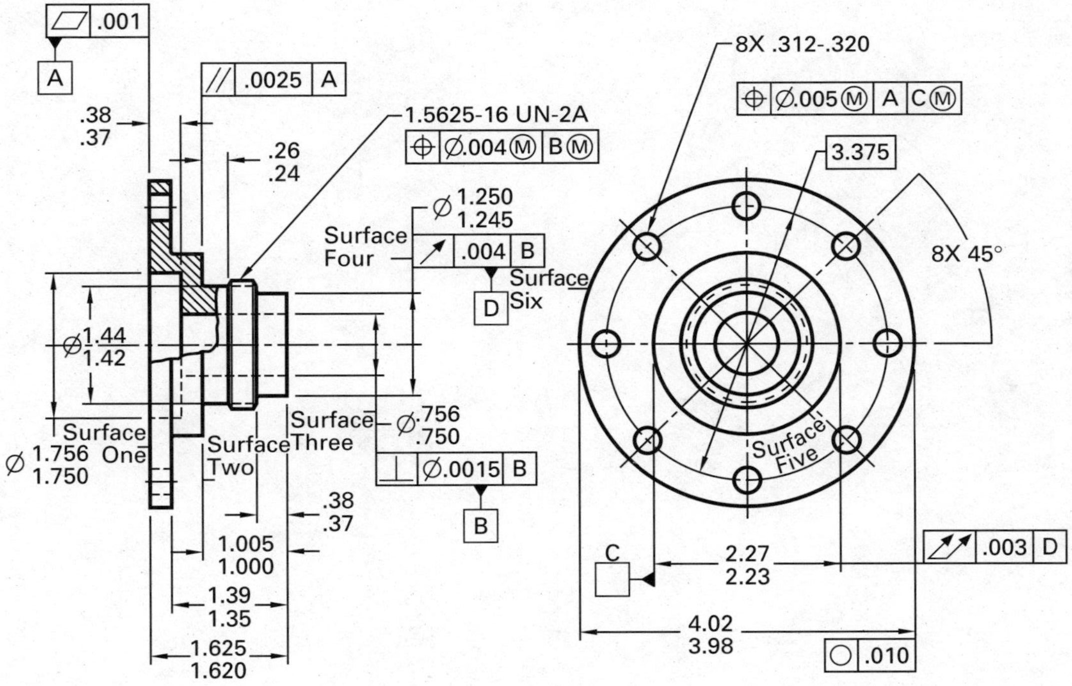

**37.** (5 points.)

What are the virtual conditions of clearance holes in the drawing of a cylinder head to be installed over ⌀.375 × 1.25-high studs in the mating part when the position tolerance for studs and holes is ⌀.020 at MMC, relative to the front and one side edge? Show all calculations.  ⌀.415

FIXED FASTENERS:

$T = \frac{H-F}{2}$   $H - F = 2T$   $H = 2T + F$
$\qquad\qquad\qquad\qquad\qquad\qquad = 2(.020) + .375$
$\qquad\qquad\qquad\qquad\qquad\qquad = .040 + .375$
$\qquad\qquad\qquad\qquad\qquad\qquad H = .415$

**38.** (8 points.)

Draw the feature control frame for the clearance holes in Problem 38.

⌖ | ⌀.020 Ⓜ 1.25 Ⓟ | A | B

**39.** (10 points.)

A batch of cover plates, shown below, was inspected, and the holes were found to be well within the position tolerance but undersize by .25 mm; therefore not acceptable. However, the inspector, who understood the concept of zero position tolerance at MMC, suggested a change in the feature control frame on the drawing, which would make the parts acceptable without rework. Draw in the space below the revised feature control frame and hole callout suggested by the inspector.

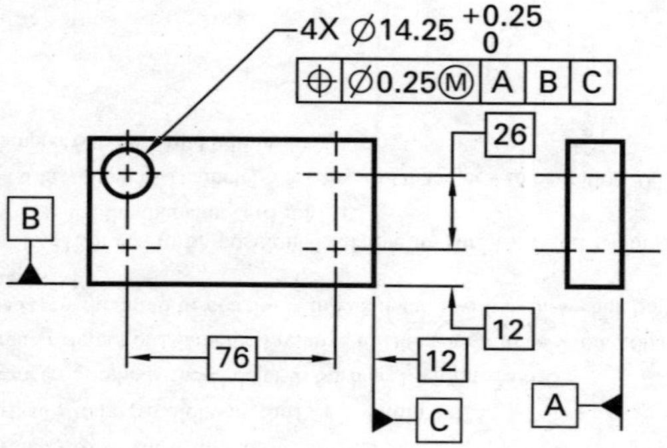

4X ⌀14.25 +0.25/0  | ⊕ | ⌀0.25 Ⓜ | A | B | C |

# Answers to Chapter Review Problems

| Chapter | Answer Page | Chapter | Answer Page |
|---------|-------------|---------|-------------|
| 1 | 61 | 11 | 72 |
| 2 | 61 | 12 | 74 |
| 3 | 62 | 13 | 75 |
| 4 | 63 | 14 | 76 |
| 5 | 65 | 15 | 78 |
| 6 | 65 | 16 | 79 |
| 7 | 67 | 17 | 80 |
| 8 | 68 | 18 | 82 |
| 9 | 70 | 19 | 85 |
| 10 | 71 | 20 | 87 |

# Chapter 1 Review Problems

Write the letter of the best answer in the space at left.

__A__   1. If geometric tolerances are not specified on a drawing, what controls the geometry?
   (a) dimensions   (b) views   (c) material   (d) notes

__C__   2. The Y14.5M standard recommends geometric tolerances be specified by the use of _____.
   (a) local notes   (b) general notes
   (c) symbols   (d) size tolerances

__C__   3. If the upper surface of a block is located by a ± .01 size tolerance from a bottom surface that is absolutely flat, the upper surface can be curved by a maximum of _____.
   (a) .005   (b) .01   (c) .02   (d) more than .02

__B__   4. When did the British start using geometric tolerancing symbols?
   (a) 1920s   (b) 1930s   (c) 1940s   (d) 1950s

__D__   5. What is the abbreviation of the name of the organization that currently governs the development of geometric tolerancing practices?
   (a) ASA   (b) USASI   (c) NASA   (d) ANSI

__C__   6. What is the year that the current Y14.5 standard was made available to the public?
   (a) 1973   (b) 1982   (c) 1994   (d) 1995

# Chapter 2 Review Problems

Next to each geometric characteristic listed below, select and sketch the appropriate symbol from the symbols shown.

1. Straightness
2. Circularity
3. Profile of a line
4. Parallelism
5. Angularity
6. Concentricity
7. Position
8. Flatness
9. Cylindricity
10. Profile of a surface
11. Perpendicularity
12. Circular runout
13. Total runout
14. Symmetry

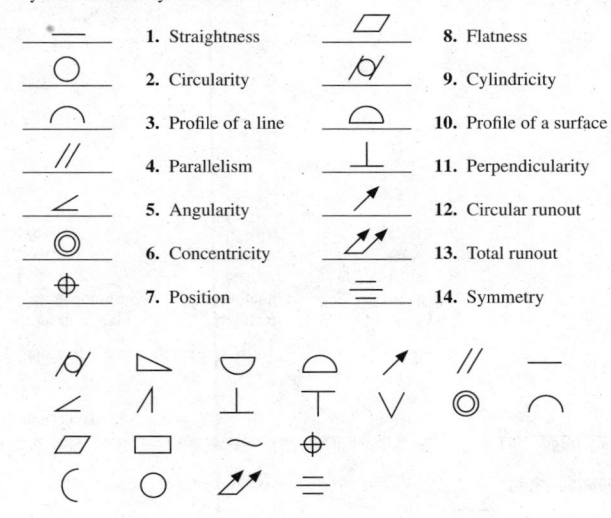

15. The symbol for *diameter* is one of those shown below. Sketch the correct symbol in the space at left.

Write the letter of the best answer in the space at left.

__C__   16. Which of the following types of lines cannot have a feature control frame attached to it?
   (a) extension   (b) leader   (c) center   (d) dimension

# Chapter 3 Review Problems, Part I

Write the letter of the best answer in the space at left.

__D__ 1. What is the name of the area of the total permissible error of a geometric tolerance?
 (a) virtual condition  (b) basic dimension
 (c) datum  (d) tolerance zone

__A__ 2. What is the maximum material condition of a hole?
 (a) smallest size  (b) largest size
 (c) nominal size  (d) basic size

__B__ 3. What is the maximum material condition of a shaft?
 (a) smallest size  (b) largest size
 (c) nominal size  (d) basic size

__D__ 4. Which of the following is the symbol for maximum material condition?
 (a) *R*  (b) *F*  (c) *S*  (d) *M*

__B__ 5. The virtual condition is the combination of the maximum material condition and the _____.
 (a) size tolerance  (b) geometric tolerance
 (c) mating size  (d) bonus tolerance

__B__ 6. What is the abbreviation of words that refer to the total movement of the probe of a dial indicator?
 (a) FIR  (b) FIM  (c) TIR  (d) TIM

__A__ 7. Which of the following is the symbol used to indicate the tolerance zone is above the surface of the part?
 (a) *P*  (b) *M*  (c) *L*  (d) *E*

__B__ 8. What is the name of the theoretically exact surface from which geometric dimensions are measured?
 (a) feature  (b) datum
 (c) datum feature  (d) nominal surface

__A__ 9. What is the name of a line, real or imaginary, that can be drawn on a surface?
 (a) element  (b) feature  (c) datum  (d) radial line

__D__ 10. What is the name of a line that extends toward or away from a center?
 (a) element  (b) center line
 (c) extension line  (d) radial line

# Chapter 3 Review Problems, Part II

In the space to the right of each statement, write the word or term from the list below that pertains to the statement.

1. Any line, real or imaginary, that can be drawn on a surface. — _Element_

2. Any portion of an object, such as a point, axis, plane, or cylindrical surface, or a tab, recess, or groove. — _Feature_

3. Describes a dimension that does not have a tolerance. The tolerance is given elsewhere. — _Basic_

4. The size of a shaft as produced is such that it contains the most material possible. — _MMC_

5. A theoretically exact surface or line from which dimensions or geometric tolerances can be taken. — _Datum_

6. The geometric tolerance specified applies, no matter how big or small the feature can be produced. — _RFS_

7. The physical state of an object that results in the tightest fit with a mating part. — _Virtual condition_

8. The total reading of a dial indicator. — _FIM_

9. Pointing toward or away from a center. — _Radial line_

10. The area taken up by the total amount of permissible geometric error. — _Tolerance zone_

Select answers from the following:

| Tolerance zone | Element | Normality |
|---|---|---|
| Virtual condition | Radial line | MMC |
| Datum | Feature | FIM |
| Basic | Characteristic | RFS |
| Median | Modifier | LMC |

62 **Answers to Review Problems**  © Glencoe/McGraw-Hill

# Test 4: Chapters 15 and 16
## Concentricity and Runout

This is a short-answer recall test in which the student is expected to write a statement or complete a sentence in the blank space. When the required answer is one word, the blank is short. For two words there are two short blanks. When a statement is required, the blank is about as long as the required answer; however, the answer does not have to be that length to be correct.

1. Two cylinders are coaxial when they have the same _____axis_____.

2. There is no symbol for the general case of coaxiality. Draw the symbols that are used for three special cases.
   ◎  ↗  ⊕

3. Concentricity error is the amount by which the axes of two regular solids, such as cylinders, are _____out of line_____.

4. Give one reason why concentricity is not as often specified as runout and position tolerance. _____It is more restrictive. It is difficult to measure._____

5. The tolerance zone for concentricity is always an imaginary cylinder. How is this fact shown in the feature control frame? _____The diameter symbol (⌀) is placed before the tolerance._____

6. If the concentricity of a cylinder is ⌀.03 mm, what is the maximum slant, in millimeters, of the feature axis relative to the datum axis? _____0.03_____

7. Concentricity does not include errors in surface characteristics. Name *one* of these characteristics. _____roundness     straightness     taper_____

8. Give an example of a machine part in which concentricity should be controlled. _____A motor rotor_____

9. Runout is any deviation of a surface from perfect form that can be detected by _____rotating the part about an axis_____.

10. Runout is a composite tolerance, including errors in other geometric characteristics. Name *one*. _____Circularity     Straightness     Perpendicularity_____

11. In circular runout, the deviation of each _____circular_____ _____element_____ is controlled.

12. Total runout controls the deviation of all _____elements_____ in a surface, circular and straight.

13. Why should circular runout be specified, rather than total runout, whenever it is adequate? _____It is less expensive to measure_____

14. Two separated cylinders on a crankshaft are to be used as one datum for a total runout tolerance of .02 mm. Draw the required feature control frame.

    | ↗ | 0.02 | A-B |

15. When is it necessary to specify a flat face as a datum in addition to a cylindrical surface? _____When the part is large in diameter and short in length_____

16. Why is it necessary to place a stop against the opposite face of the part when inspecting a flat face for runout? _____To prevent axial movement_____

17. A(n) _____datum_____ is required with concentricity, runout, and position tolerance because these are relationship features.

18. What material modifier applies to concentricity and runout, unless otherwise specified? _____Regardless of feature size_____

19. A high-speed turbine rotor shaft must maintain very close coaxiality in order to minimize imbalance. The designer decides this is an RFS application, but which coaxiality tolerance should be specified? Why? _____Concentricity-coaxiality must be maintained independently of surface errors_____

# Test 5: Chapters 17–20
## Position Tolerance—General and Location Applications
## Position-Coaxial Applications and Symmetry

This is a short-answer recall test in which the student is expected to write a statement or complete a sentence in the blank space. When the required answer is one word, the blank is short. For two words there are two short blanks. When a statement is required, the blank is about as long as the required answer; however, the answer does not have to be that length to be correct.

All questions have a grading point value of one unless otherwise specified. Total points = 30.

1. Position *error* is any __deviation (variation)__ in the location of a feature.

   Position *tolerance* is the total __permissible error (variation)__ in the location of a feature. (2 points)

2. Like runout, position tolerance is a composite tolerance, including surface errors in addition to location error. Name *two* of the possible surface errors.

   __Circularity__   __Straightness__

3. For a feature located by its axis, such as a round hole, the shape of the position tolerance zone is an imaginary __cylinder__.

4. For a feature located by its center plane, such as a rectangular slot, the position tolerance zone is the space between two imaginary __planes__.

5. Coordinate dimensioning results in a square or rectangular tolerance zone in which the permissible error is not the same in all directions. This is corrected in position tolerance, since the tolerance zone is __circular (round)__ in shape.

6. A basic dimension is an exact value with no specified __tolerance__.

7. The __tolerance__ for a basic dimension is given in the __feature control frame__.

8. What type of tolerance is often controlled when using position, in addition to location? __Orientation tolerances__

9. The LMC modifier is most often used to control a critical __edge distance__.

10. How does an RFS modifier affect the specified position tolerance? __No effect__

11. In a floating fastener assembly, what is allowed to float relative to what? __The bolts in the clearance holes__

12. A projected tolerance zone transfers the position tolerance from the part being controlled to the __mating__ __part__.

13. Where the fasteners are studs or press-fit pins, the projected tolerance zone must be equal not to the thickness of one of the parts but to the __height of the stud or pin__.

14. When a position tolerance specification is converted to zero tolerance at MMC, the position tolerance is reduced to zero, but this is compensated for by increasing the __size tolerance of the hole__.

15. In designing the fit of a tab in a slot, the total tolerance in both parts is made equal to the minimum clearance between the mating features. Write the formula used for the calculation.  $T_{TOT} = W_S - W_T$

16. Position tolerance is used in place of angularity for a size feature when what modifier is applied? __MMC__

17. Symmetry error is the amount by which opposite sides of a size feature are unequally spaced from the __center__ __plane__ of the datum.

18. Two or more feature patterns on the same part are called __multiple patterns of features__.

19. Multiple hole patterns located by basic dimensions from the same datums are considered one __composite__ __pattern__.

20. How many gages are used to inspect a composite hole pattern? __One__

21. A plate has three different hole patterns, and each one is to be gaged separately. What note is added below the feature control frame? __Sep Reqt__

22. The form of the tolerance zone for coaxial position tolerance is an imaginary __cylinder__.

# Test 1Q: Chapters 1–6
## Introduction, Symbols, Terms, Datums, Inspection of Geometric Tolerances, General Rules

This is a short-answer recall test in which the student is expected to write a statement or complete a sentence in the blank space. When the required answer is one word, the blank is short. For two words there are two short blanks. When a statement is required, the blank is about as long as the required answer; however, the answer does not have to be that length to be correct.

1. How was geometric tolerance specified on drawings before the introduction of the symbol system? __*By local notes*__

2. List six geometric characteristics and draw the correct symbol next to each. __*(Any six characteristics of text Fig. 2–1)*__

3. Draw full size an example of a datum feature symbol (also called a datum identifying symbol).  [A]

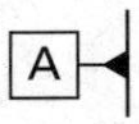

4. There are four modifier symbols. Draw the symbols for *three* of them.  Ⓜ Ⓛ Ⓟ

5. Make a sketch of a simple bar and add a feature control frame for straightness with a tolerance of .002. Omit dimensions.    [— | .002]

6. An element may be a straight line or it may be a(n) __*circle*__.

7. How does a datum *plane* differ from a datum *feature*? __*The datum plane has perfect form; the datum feature is an actual imperfect surface*__

8. The size of an object that results in the tightest fit with the mating part is called the __*virtual*__ condition.

9. On drawings that do not specify geometric tolerances, what determines the geometric form? __*The size tolerances*__

23. In situations where maintaining a minimum wall thickness is more critical than coaxiality, which modifier is specified? __*LMC*__

24. An engine block has five coaxial holes to support a camshaft. The common axis must be positioned relative to three surfaces of the block within .005 diameter, while the axes of the individual holes must be all within .0005 diameter, completely within the .005 diameter. Draw a composite feature control frame to specify this design intent.

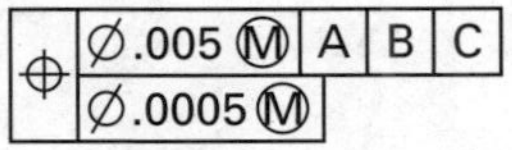

   ⊕ | ⌀.005 Ⓜ | A | B | C
   ⊕ | ⌀.0005 Ⓜ

25. In order to avoid interference in fitting coaxial parts, the total position tolerance for both parts must not be greater than the __*minimum*__ __*clearance*__ between the parts.

26. When a coaxiality tolerance for a bore is specified as zero at MMC, where does the machinist find the actual manufacturing tolerance? __*In the bore size tolerance*__

27. Two parts having mating coaxial features fit together. Which type of coaxial control should the designer specify on the two detail drawings? __*Position*__

# Test 2Q: Chapters 7–11
## Straightness, Flatness, Circularity, Cylindricity, Profile

This is a short-answer recall test in which the student is expected to write a statement or complete a sentence in the blank space. When the required answer is one word, the blank is short. For two words there are two short blanks. When a statement is required, the blank is about as long as the required answer; however, the answer does not have to be that length to be correct.

1. How does flatness differ from straightness? *Flatness controls all the elements in all*

   *directions*

2. Circularity error is the measure of how much each circular element of an object deviates from being a(n) _*circle*_

3. How does cylindricity differ from circularity? *Cylindricity controls straight elements*

   *as well as circular elements*

4. Draw the correct symbol for each geometric characteristic. Circularity ○

   Straightness —   Flatness ▱   Cylindricity ⌭

5. Which modifier applies to straightness unless otherwise specified? _*RFS*_

6. Draw an example of a feature control frame for a unit straightness of .001 per inch of length.

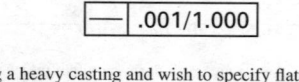

   | — | .001/1.000 |

7. You are designing a heavy casting and wish to specify flatness only in a .50 × 1.00 area of a large surface. Make a sketch below of the controlled area and specify a flatness tolerance of .001. Omit dimensions.

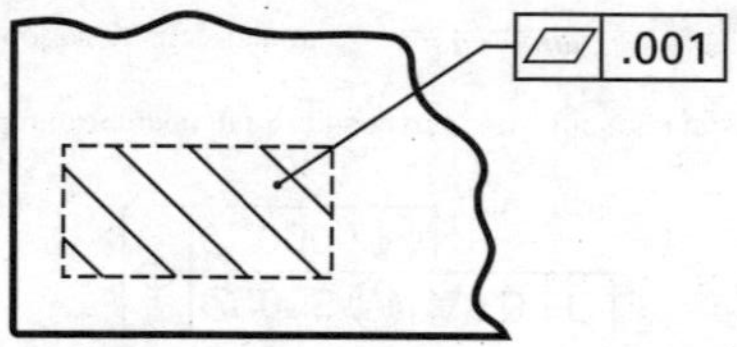

   | ▱ | .001 |

8. Cylindricity is a combination of which two geometric characteristics?

   *Circularity and straightness*

9. In profile tolerances, the tolerance is assumed to be bilateral, unless

   *otherwise specified*

10. An automotive valve cam has a profile of a surface tolerance of .05 mm that applies all around. No datum is necessary. Draw an appropriate feature control frame.

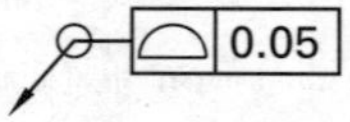

    ⌒ | 0.05

---

50   Periodic Test 2Q, Answers © Glencoe/McGraw-Hill

## Test 3Q: Chapters 12–16
## Parallelism, Perpendicularity, Angularity, Concentricity, and Runout

This is a short-answer recall test in which the student is expected to write a statement or complete a sentence in the blank space. When the required answer is one word, the blank is short. For two words there are two short blanks. When a statement is required, the blank is about as long as the required answer; however, the answer does not have to be that length to be correct.

1. When are two lines parallel? *When every point on one line is the same distance from opposing points on another line*

2. When a parallelism tolerance zone is an imaginary cylinder, the tolerance is specified as a diameter. With which type of feature does this occur? *A cylinder*

3. Perpendicularity is a special case of which other geometric characteristic? *Angularity*

4. The angle given in an angularity tolerance is basic (no tolerance), but some manufacturing error must be allowed. Where is the tolerance specified? *In the feature control frame*

5. Concentricity tolerance is the maximum permissible error in *coaxiality (alignment of the axes)*

6. If the concentricity of a cylinder is ⌀0.03 mm, what is the maximum slant, in millimeters, of the feature axis relative to the datum axis? *.03*

7. Runout is any deviation of a surface from perfect form that can be detected by *rotating the part about an axis*

8. Total runout is a composite tolerance, including errors in other geometric characteristics. Name *one*. *Circularity, straightness, taper, perpendicularity*

## Test 4Q: Chapters 17 and 18
## Position Tolerance—General and Location Applications

This is a short-answer recall test in which the student is expected to write a statement or complete a sentence in the blank space. When the required answer is one word, the blank is short. For two words there are two short blanks. When a statement is required, the blank is about as long as the required answer; however, the answer does not have to be that length to be correct.

1. Position error is any *deviation (variation)* in the location of a feature.

    Position *tolerance* is the total *permissible error (variation, deviation)* in the location of a feature. (2 points)

2. Coordinate dimensioning results in a square or rectangular tolerance zone in which the permissible error is not the same in all directions. This is corrected in position tolerance, since the tolerance zone is *circular (round)* in shape.

3. In a floating fastener assembly, what is allowed to float relative to what? *The bolts in the clearance holes*

4. A basic dimension is an exact value with no specified *tolerance*

5. When a position tolerance specification is converted to zero tolerance at MMC, the position tolerance is reduced to zero, but this is compensated for by increasing the *size tolerance of the hole*

6. In designing the fit of a tab in a slot, the total tolerance in both parts is made equal to the minimum clearance between the mating features. Write the formula used for the calculation. $T_{TOT} = W_S - W_T$

7. In addition to location, what type of tolerance is often controlled when using position? *Orientation*

© Glencoe/McGraw-Hill

# Answers to Final Examination

Student tests, pages 31–39.

# Chapter 4 Review Problems, Part I

Write the letter of the best answer in the space at left.

__B__ 1. Which of the following would be a proper datum?
 (a) lathe center  (b) bearing surface
 (c) center line  (d) axis

__A__ 2. Which of the following groups of letters are not used by themselves on an engineering drawing?
 (a) I, O, Q  (b) A, B, C  (c) M, S, P  (d) X, Y, Z

__A__ 3. What can be a datum feature?
 (a) feature of a part shown on a drawing
 (b) feature of the tooling used to manufacture the part
 (c) neither (a) nor (b)
 (d) both (a) and (b)

__D__ 4. Where does a datum plane exist?
 (a) on the drawing  (b) on the part
 (c) in the tooling  (d) in theory only

__D__ 5. What is a simulated datum?
 (a) machine tool mounting surface
 (b) assembly fixture mounting surface
 (c) inspection fixture mounting surface
 (d) any of the above

__B__ 6. Which of the following is not a reason to use a datum reference frame?
 (a) restrict degrees of freedom
 (b) move parts between locations
 (c) repeat location of parts
 (d) establish measurement surfaces

__D__ 7. Datum features should be selected because _____.
 (a) they are functional  (b) they are mating features
 (c) they are readily accessible  (d) all of the above

__B__ 8. The MMC size of a hole is when the hole is _____.
 (a) as large as it can be  (b) as small as it can be
 (c) its nominal size  (d) none of the above

__A__ 9. Determining the condition of a size datum is primarily based on _____.
 (a) the relationship between mating parts
 (b) the type of material
 (c) the size of the feature
 (d) whether the feature is internal or external

__D__ 10. A datum target can be a _____.
 (a) point  (b) line  (c) surface area  (d) any of the above

__A__ 11. Where are the datum target identifying letter and number placed in the datum target symbol?
 (a) in the lower half  (b) in the upper half
 (c) above the symbol  (d) below the symbol

__C__ 12. What is the name for a datum that consists of two or more individual surfaces used to restrict one principal degree of freedom?
 (a) combined datum  (b) multiple datum
 (c) compound datum  (d) equalizing datum

__D__ 13. What is the name for a datum used to center a noncircular part?
 (a) combined datum  (b) multiple datum
 (c) compound datum  (d) equalizing datum

## Chapter 4 Review Problems, Part II

1. On the isometric drawing below of an unsupported object in space, select suitable datum targets (tooling points) on the three visible surfaces to make a complete datum frame. Show the point locations with crosses. Omit dimensions.

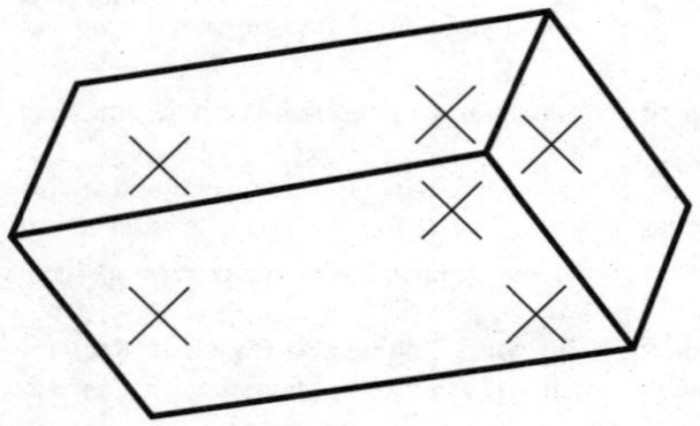

2. Add datum target symbols to both views in the drawing below, complete with datum identifying letters and target numbers.

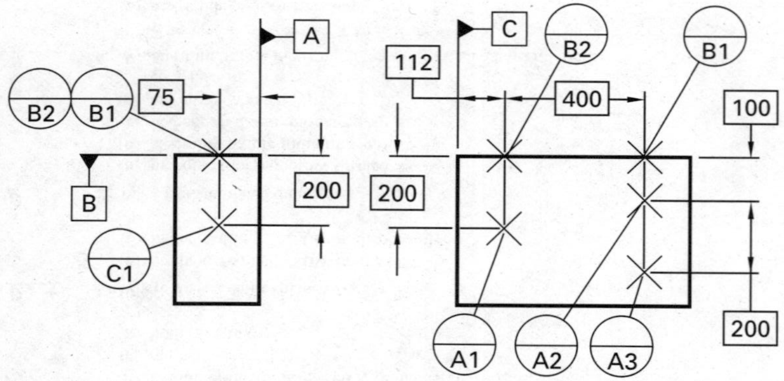

3. The cylindrical part below is mounted between two bearings on the small diameters on each end. Make each small diameter and the left end face separate datums.

4. Add a complete datum frame to the cast tube below. Select suitable datum targets on cast surfaces. Label the datum planes, give location dimensions for all datum targets, and show all datum target symbols.

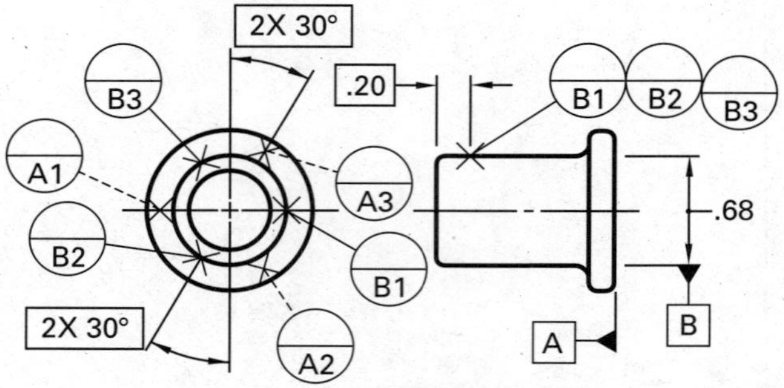

5. In the partial full-size view below, two of three datum targets for the first datum are shown. They are intended to be on the far side (not visible). Add appropriate location dimensions, datum target symbols, and leaders for the two datum targets only. The round end of the object is intended to be supported by a 90° vee-block, which will be the second datum. Add the necessary datum targets, location dimensions, and datum target symbols.

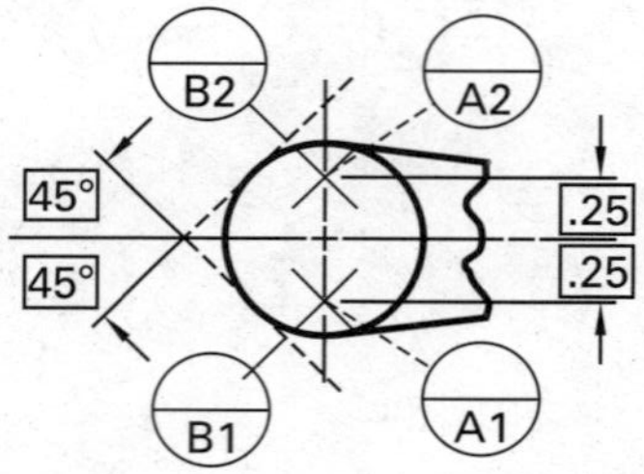

# Chapter 5 Review Problems

Write the letter of the best answer in the space at left.

__B__ 1. If a dial indicator is used to check parallelism and the FIM is .004, what is the minimum parallelism error?
   (a) .002   (b) .004   (c) .005   (d) .008

__D__ 2. What is the name of the part on a dial indicator that rotates?
   (a) clamp   (b) case   (c) probe   (d) bezel

__C__ 3. If the movement observed on a dial indicator is .001 to the left and .003 to the right, what is the FIM?
   (a) .001   (b) .003   (c) .004   (d) .008

__D__ 4. What does the "F" in FIM stand for?
   (a) free   (b) fine   (c) final   (d) full

__A__ 5. What lot size would most likely be used for measuring with a dial indicator or coordinate measuring machine?
   (a) 10   (b) 1000   (c) 5000   (d) 10,000

__A__ 6. A functional gage used for inspecting a machined part simulates the fit of the _____
   (a) mating part   (b) dial indicator
   (c) surface plate   (d) vee-block

__C__ 7. In what direction does the pointer of a dial indicator move?
   (a) clockwise only
   (b) counterclockwise only
   (c) clockwise and counterclockwise
   (d) radially

__B__ 8. What is the name of the CMM inspection surface?
   (a) surface table   (b) surface plate
   (c) measuring surface   (d) none of the above

# Chapter 6 Review Problems, Part I

Write the letter of the best answer in the space at left.

__D__ 1. What type of tolerance controls the form of a part as well as the size when no geometric tolerance is specified?
   (a) internal   (b) external   (c) virtual condition   (d) size

__A__ 2. When a geometric tolerance is applied to a size feature, which condition applies automatically?
   (a) RFS   (b) MMC   (c) LMC   (d) virtual condition

__D__ 3. When a geometric tolerance is applied to a size feature referenced to a geometrically controlled size datum, which condition applies to the datum feature?
   (a) RFS   (b) MMC   (c) LMC   (d) virtual condition

__C__ 4. Unless otherwise stated, a geometric tolerance applies to which part of a screw thread?
   (a) major diameter   (b) minor diameter
   (c) pitch diameter   (d) none of the above

__B__ 5. The limits of size rule states that no element of a part shall extend beyond what boundary?
   (a) RFS   (b) MMC   (c) LMC   (d) virtual condition

__A__ 6. The limits of size rule applies only to
   (a) individual features   (b) interrelated features
   (c) bar stock   (d) twist drills

__D__ 7. The envelope principle makes possible the calculation of a(n)
   (a) allowance   (b) maximum clearance
   (c) position tolerance   (d) (a) and (b)

# Chapter 6 Review Problems, Part II

In the space below, draw the end view of a bar, full size, that is 49.05–50.05 mm wide (left to right) and 12.00–12.05 mm high. Add limit dimensions.

1. On a bar produced from this drawing, what will be the sizes of the height and width at the MMC boundary?

   _12.05_ × _50.05_

2. When the part drawn above is produced at its MMC boundary, what is the permissible width and height error in its geometric form?

   _none_

3. In the straightness tolerance shown below, no modifier is specified. Which modifier applies automatically?

   | — | .005 |

   _RFS_

4. In the space below, draw the feature control frame for a screw thread whose pitch diameter is to be positioned to datum A within .005 inches. No screw thread drawing is necessary.

   | ⊕ | ⌀.005 | A |

5. In the space below, draw the feature control frame to control the circular runout of the pitch diameter of a gear to datum C within .003 inches. No gear drawing is necessary.

   | ↗ | .003 | C |

6. In the drawing below, the four holes are positioned relative to datums A and B within .010 inch when the holes and datum A are at MMC. Datum A is a size feature with its own position specification. What is the virtual condition of datum A?

   _.997_

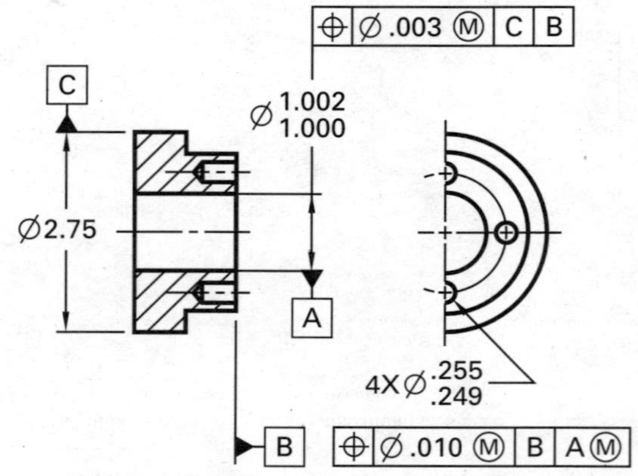

   $VC = MMC - GT$
   $= 1.000 - .003$
   $= .997$

# Chapter 7 Review Problems, Part I

Write the letter of the best answer in the space at left.

__B__  1. What is controlled when straightness is specified?
   (a) surface   (b) element   (c) chord   (d) datum

__A__  2. What is the configuration of the tolerance zone for straightness applied to a plane feature?
   (a) parallel lines   (b) cylindrical   (c) square   (d) conical

__C__  3. How is straightness applied unless otherwise specified?
   (a) at MMC   (b) at LMC
   (c) RFS   (d) to a limited length

__C__  4. Straightness of a size feature is the only control that will allow the feature to exceed _____.
   (a) LMC   (b) virtual condition
   (c) MMC   (d) basic dimension

__B__  5. What is straightness called when it is applied for a certain distance?
   (a) digital   (b) unit   (c) dimensional   (d) partial

__A__  6. What is the configuration of the tolerance zone for straightness applied to a cylinder?
   (a) cylinder   (b) cone   (c) ring
   (d) space between two parallel straight lines

__D__  7. In which view is straightness of a flat surface applied?
   (a) front view   (b) side view   (c) where the surface is an area
   (d) where the surface appears as an edge

# Chapter 7 Review Problems, Part II

1. In the space below, draw the feature control frame for a straightness tolerance of .002 inch applied to a surface.

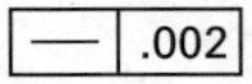

2. In the space below, draw two views (front and end) of a part that is 2 inches wide (left to right), .75 inch high, and .50 inch deep. Omit dimensions. Add a .005-inch element straightness tolerance to the top surface in the front view and a .002-inch element straightness tolerance to the rear surface.

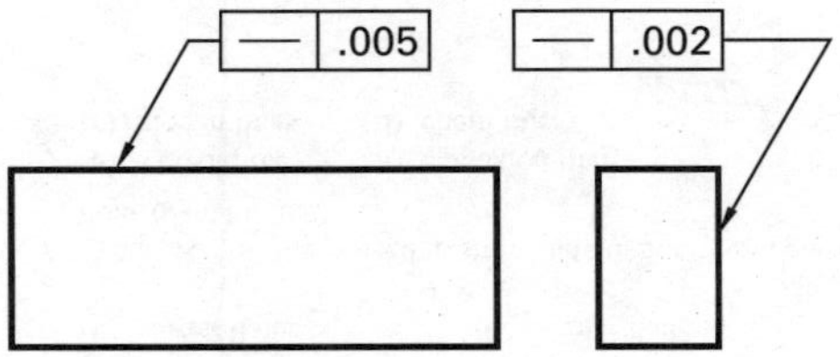

3. In the space below, repeat the front view of the drawing in problem 2 and add a two-place height dimension. Show with a properly placed feature control frame that the height is to be straight within .003 inch.

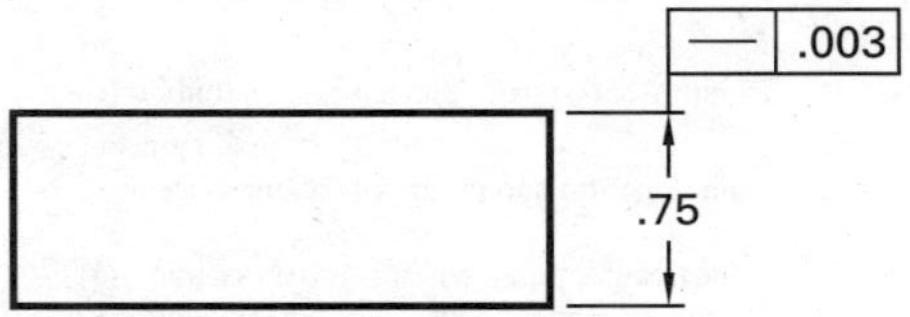

# Chapter 8 Review Problems, Part I

Write the letter of the best answer in the space at left.

__A__ 1. A flatness tolerance zone is the space between two parallel _____.
   (a) planes   (b) elements   (c) straight lines   (d) datums

__D__ 2. To which of the following views of a surface is a flatness feature control frame directed?
   (a) true size   (b) horizontal
   (c) vertical   (d) an edge view

__C__ 3. What is the measure of how much a plane surface deviates from being a true plane?
   (a) plane deviation   (b) plane variation
   (c) flatness error   (d) a tolerance zone

__C__ 4. A surface can be straight in one direction but not be flat. The preceding statement _____.
   (a) requires more data   (b) is questionable
   (c) is true   (d) is false

__B__ 5. No element of a surface controlled by a flatness tolerance may extend beyond what boundary?
   (a) LMC   (b) MMC   (c) RFS   (d) virtual condition

__B__ 6. What is flatness specified per area called?
   (a) dimensional flatness   (b) unit flatness
   (c) partial flatness   (d) digital flatness

__D__ 7. What type of line is used to show the boundary of a particular area controlled by flat?
   (a) a break line   (b) extension lines
   (c) hidden lines   (d) chain lines

4. A long round shaft has a unit size straightness of .001 inch per 1.000-inch length regardless of feature size. Draw the appropriate feature control frame below.

| — | ⌀.001/1.000 |

5. A long round shaft has a unit size straightness of .001 inch per 1.000-inch length at MMC, but the total straightness at MMC must be within .010 inch. Draw the appropriate feature control frame below.

| — | ⌀.010 Ⓜ |
| — | ⌀.001/1.000 Ⓜ |

68   Answers to Review Problems                                  © Glencoe/McGraw-Hill

# Chapter 8 Review Problems, Part II

**1.** In the space below, draw the feature control frame for a flatness tolerance of .002 inch.

⌗ .002

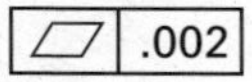

**2.** In the space below, draw two views (front and end) of a part that is 2 inches wide (left to right), .75 inch high, and .50 inch deep. Omit dimensions. Add a flatness tolerance of .005 inch to the top surface and a flatness tolerance of .002 inch to the rear surface.

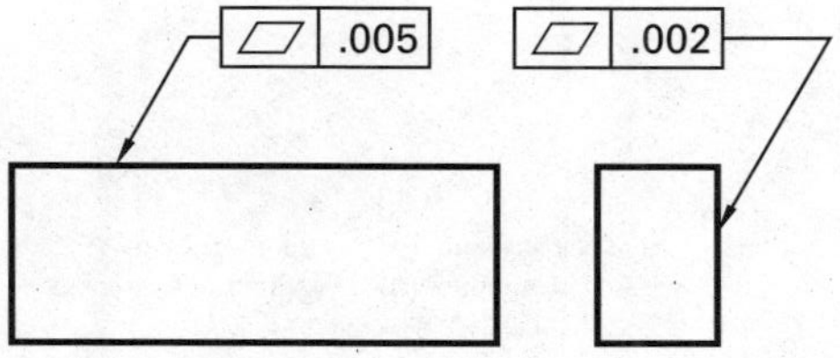

⌗ .005      ⌗ .002

**3.** In the drawing below, with no flatness control specified, what is the maximum permissible flatness error of the top surface if the bottom surface is perfectly flat?

*.004*

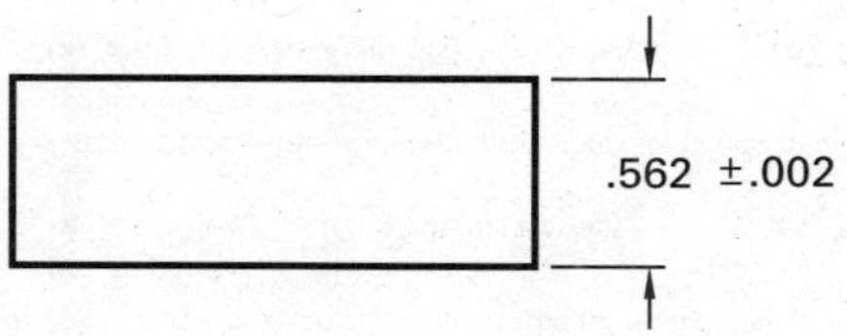

.562 ±.002

**4.** In the drawing for problem 3, what is the maximum permissible flatness error of the top surface if the bottom surface is out-of-flat by .001 inch?

*.003*

**5.** In the space below, redraw the drawing for problem 3 and add flatness tolerances of .001 inch to the top and bottom surfaces.

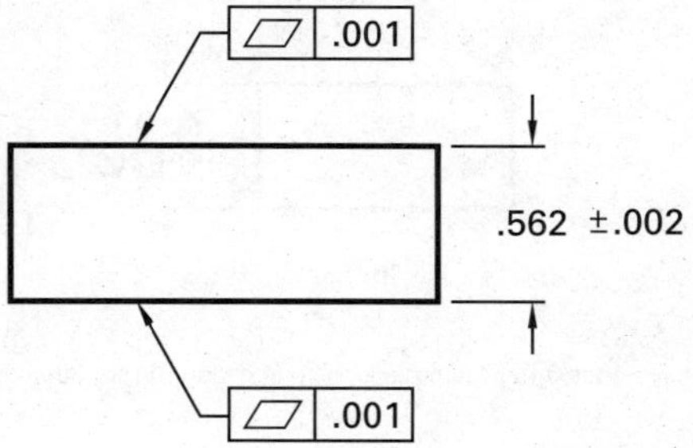

⌗ .001

.562 ±.002

⌗ .001

**6.** Shown below is one possible enlarged interpretation of the student's drawing for problem 5. Given the size and flatness tolerances of the correct drawing and this interpretation, fill in the values of dimensions A through D in the spaces provided.

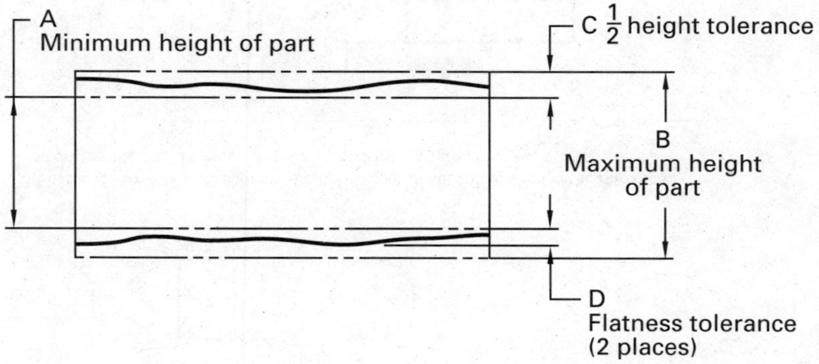

A — Minimum height of part
B — Maximum height of part
C — ½ height tolerance
D — Flatness tolerance (2 places)

A  *.560*   B  *.564*   C  *.002*
D  *.001*

# Chapter 9 Review Problems, Part I

Write the letter of the best answer in the space at left.

__C__   1. What does circularity control?
(a) features   (b) surfaces   (c) elements   (d) datums

__C__   2. Does a circularity tolerance have to be within the dimensional limits of size?
(a) it may   (b) it may not   (c) it must be   (d) it must not

__C__   3. A circularity tolerance always applies _____.
(a) FIM   (b) at LMC
(c) RFS   (d) at MMC

__D__   4. What is the most likely shape of the tolerance zone for circularity?
(a) a cylinder   (b) a radius
(c) an ellipse   (d) a ring or washer

__A__   5. Circularity error is what type of distance between two concentric perfect circles?
(a) radial   (b) diametral   (c) angular   (d) circular

__C__   6. A circularity tolerance can be what type of tolerance curled into a circle?
(a) concentricity   (b) cylindricity
(c) straightness   (d) flatness

__A__   7. For circularity to be most accurately inspected, what is a part placed on?
(a) on a turntable   (b) on lathe centers
(c) in a vee-block   (d) on a surface plate

70   **Answers to Review Problems**

# Chapter 9 Review Problems, Part II

All the dimensions in these problems are in metric units (millimeters).

1. In the space below, draw the feature control frame for a circularity tolerance of .05 mm.

◯ 0.05

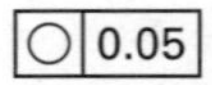

2. In the space below, draw two views of a cylinder 15 mm in diameter and 10 mm long. Space the views about 20 mm apart. Omit dimensions. Add a circularity tolerance of .05 mm.

◯ 0.05

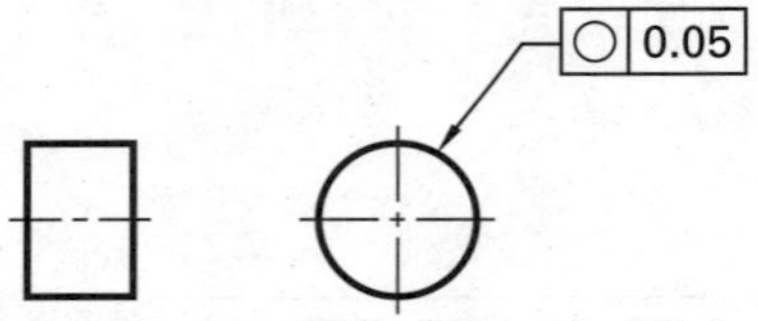

3. In the drawing below, with no circularity control specified, what is the maximum permissible circularity error of the cylindrical surface?

0.04

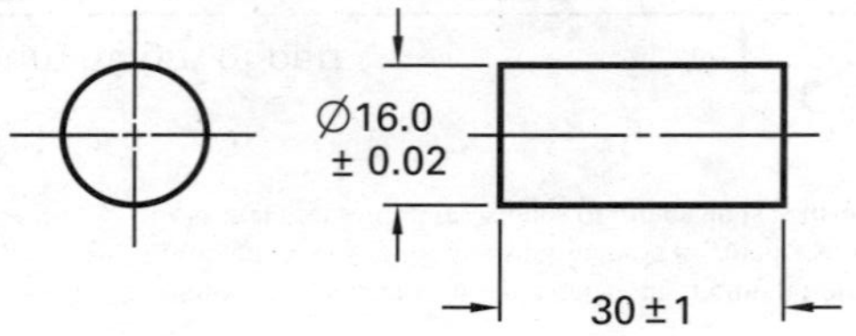

⌀16.0 ± 0.02       30 ± 1

4. In the space below, redraw the drawing for problem 3 and add a circularity tolerance of 0.01 mm to the cylindrical surface.

◯ 0.01

⌀16.0 ± 0.02       30 ± 1

© Glencoe/McGraw-Hill

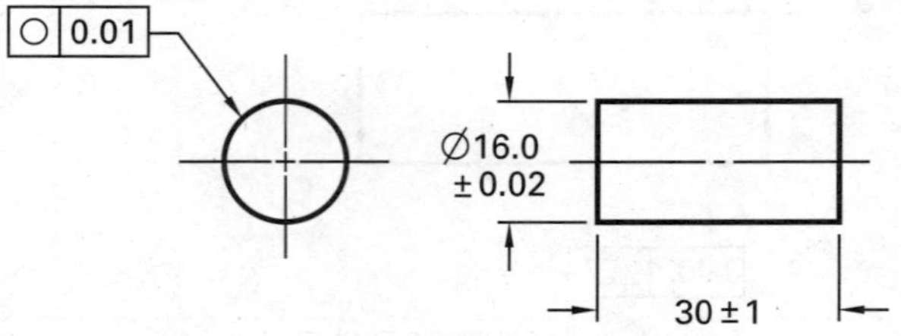

# Chapter 10 Review Problems, Part I

Write the letter of the best answer in the space at left.

_B_  1. Cylindricity is a combination of circularity and _____.
   (a) flatness  (b) straightness
   (c) parallelism  (d) perpendicularity

_A_  2. Circularity error is what type of distance between two concentric perfect circles?
   (a) radial  (b) diametral  (c) total  (d) angular

_C_  3. What is the most likely shape of the tolerance zone for cylindricity?
   (a) doughnut  (b) globe  (c) tube  (d) cylinder

_C_  4. A cylindrical tolerance _____ be within the dimensional size limits of the part.
   (a) may  (b) may not  (c) must  (d) must not

_C_  5. A cylindricity tolerance always applies _____.
   (a) FIM  (b) at LMC
   (c) RFS  (d) at MMC

_B_  6. A cylindricity tolerance can be what type of tolerance curled into a circle?
   (a) straightness  (b) flatness  (c) concentricity  (d) profile

_B_  7. What must the longitudinal elements be checked for in cylindricity?
   (a) concentricity  (b) straightness
   (c) flatness  (d) coaxiality

5. Shown below is one possible enlarged interpretation of the drawing for problem 4. Given the size and circularity tolerance of the correct drawing and this interpretation, fill in the values of dimensions A through D in the spaces provided.

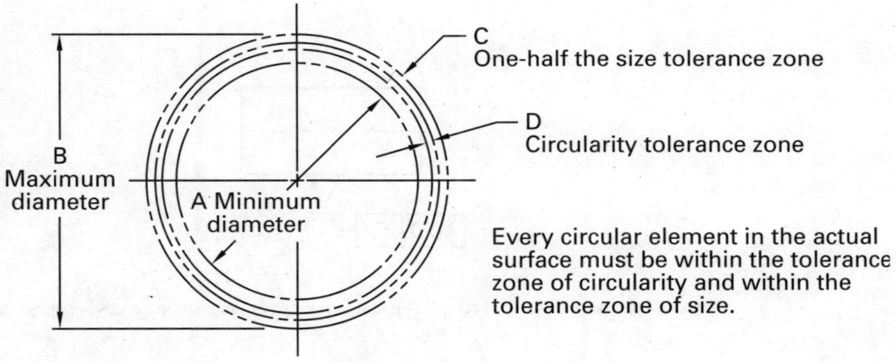

C — One-half the size tolerance zone

D — Circularity tolerance zone

A Minimum diameter

B Maximum diameter

Every circular element in the actual surface must be within the tolerance zone of circularity and within the tolerance zone of size.

A  _15.98_   B  _16.02_   C  _0.02_
D  _0.01_

© Glencoe/McGraw-Hill

Answers to Review Problems  71

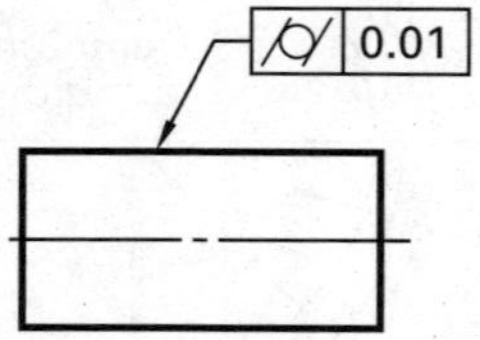

## Chapter 11 Review Problems, Part I

Write the letter of the best answer in the space at left.

__C__ 1. What does profile of a line control?
(a) chords  (b) radii  (c) elements  (d) arcs

__B__ 2. Unless otherwise specified, how is profile assumed to be applied?
(a) unilaterally  (b) bilaterally  (c) unidirectionally  (d) aligned

__A__ 3. When a phantom line and a dimension are used to show a profile tolerance, how is it applied?
(a) unilaterally  (b) bilaterally  (c) unidirectionally  (d) aligned

__C__ 4. Profile of a surface tolerance controls the profile in _____.
(a) one direction  (b) two directions
(c) three directions  (d) one curved line

__A__ 5. Which geometric control is profile of a line most similar to?
(a) straightness  (b) flatness
(c) cylindricity  (d) parallelism

__D__ 6. What must be specified on the drawing when a profile tolerance is related to another feature of the part?
(a) unilateral tolerance  (b) bilateral tolerance
(c) position tolerance  (d) datum

__A__ 7. What is the leader directed to when profile of a surface is used to control coplanarity?
(a) an extension line between the surfaces
(b) the larger surface
(c) a dimension line locating the surfaces
(d) the surface that is to be machined first

__D__ 8. What may profile of a surface control?
(a) form  (b) orientation
(c) position  (d) all of the above

## Chapter 10 Review Problems, Part II

1. In the space below, draw the feature control frame for a cylindricity tolerance of .0005 inch.

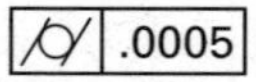

2. In the space below, draw two views of a cylinder .63 inch in diameter and 1 inch long. Space the views about 1.5 inches apart. Omit dimensions. Add a cylindricity tolerance of .003 inch.

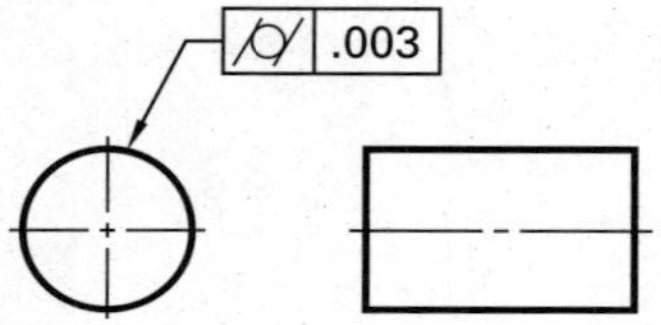

3. In the drawing below, with no cylindricity control specified, what is the maximum permissible cylindricity error?

__0.1__

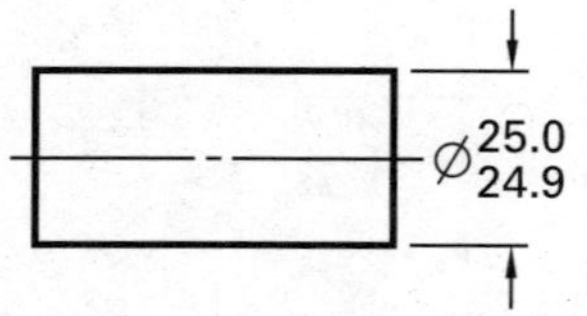

4. In the space below, redraw the drawing for problem 3 and add a cylindrical tolerance of .01 inch.

## Chapter 11 Review Problems, Part II

1. On the drawing below, show that the *curved elements* are within .1 mm total bilateral tolerance of the desired profile. The profile tolerance applies between points X and Y.

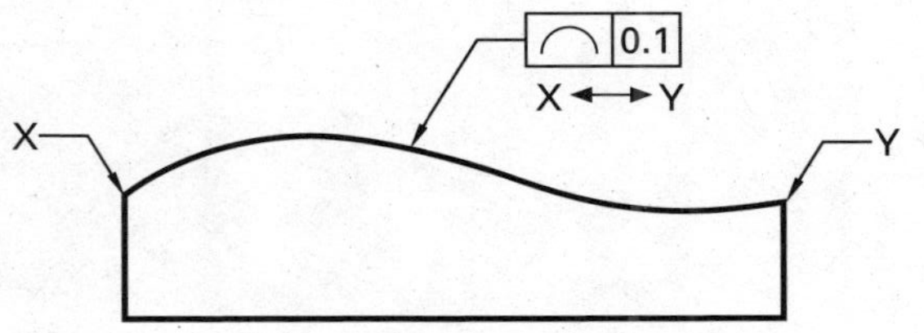

2. Redraw below the figure in problem 1 to make the tolerance only on the inside of the true profile.

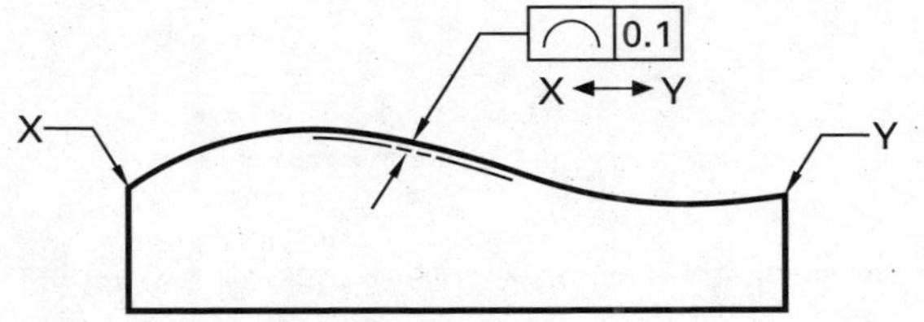

3. On the drawing below, show that the entire curved surface is to have a bilateral profile tolerance of 0.1 mm.

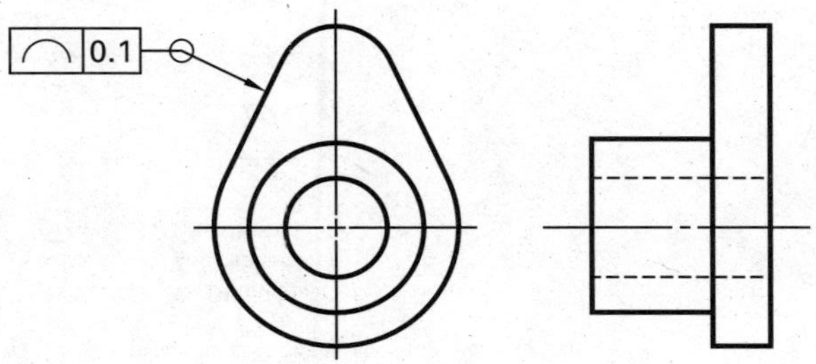

4. On the same cam, repeated below, show that the entire curved surface is to have a profile tolerance of .05 mm *outward* only of the desired profile.

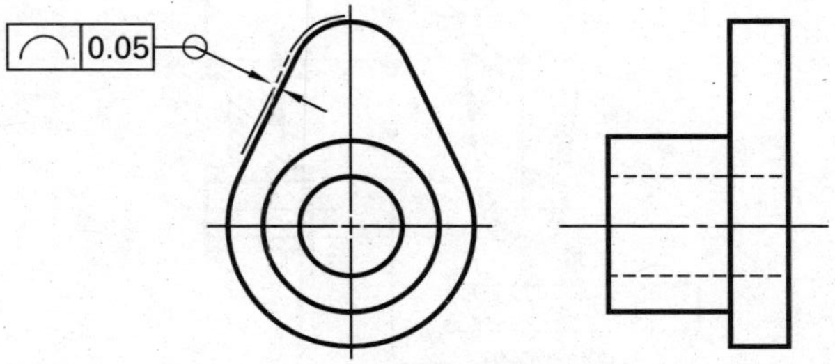

# Chapter 12 Review Problems, Part I

Write the letter of the best answer in the space at left.

__A__   1. Which of the following geometric characteristics does not require a datum?
    (a) cylindricity    (b) parallelism
    (c) perpendicularity    (d) concentricity

__B__   2. How many forms may parallelism take?
    (a) two    (b) three
    (c) four    (d) five

__C__   3. Which symbol is placed in front of the geometric tolerance when the tolerance zone has a cylindrical shape?
    (a) parallel    (b) cylindricity    (c) diameter    (d) area

__C__   4. How is a parallelism tolerance applied when no modifier is specified?
    (a) at MMC    (b) at LMC
    (c) RFS    (d) at virtual condition

__C__   5. If a surface is controlled by a .002 parallelism tolerance, what is the maximum flatness error?
    (a) .0005    (b) .001    (c) .002    (d) .004

__D__   6. When the MMC symbol is specified for a parallelism tolerance, the controlled feature is _____.
    (a) parallel to a datum    (b) as large as possible
    (c) as accurate as possible    (d) a size feature

# Chapter 12 Review Problems, Part II

1. A surface of a part is parallel with the opposite surface within .005. In the space below, draw the appropriate feature control frame.

    // | .005 | A

2. A hole in a part is parallel with the flat base of the part within .002, regardless of the produced size of the hole. In the space below, draw the appropriate feature control frame.

    // | ⌀.002 | A

3. In the drawing below, the upper hole at MMC should be parallel with the lower hole within .002 diameter, regardless of the produced size of the lower hole. Add the appropriate symbols to the drawing.

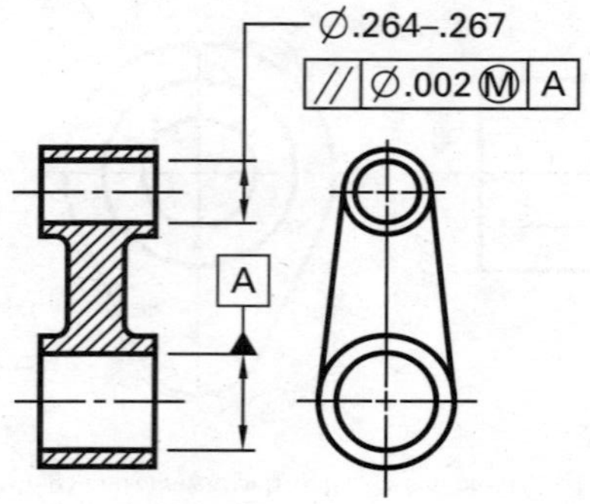

⌀.264–.267

// | ⌀.002 Ⓜ | A

4. In problem 3, the parallelism tolerance varies with the produced size of the upper hole. Complete the table below, showing this variation.

| Feature Size | Diameter Tolerance Zone Allowed |
|---|---|
| .264 | .002 |
| .265 | .003 |
| .266 | .004 |
| .267 | .005 |

# Chapter 13 Review Problems, Part I

Write the letter of the best answer in the space at left.

__C__  1. Which of the following is not a possible form for perpendicularity?
 (a) parallel lines   (b) parallel planes
 (c) parallel curves   (d) cylinder

__A__  2. What must the form of the tolerance zone be before a diameter symbol is placed in front of the geometric tolerance?
 (a) cylinder   (b) circle   (c) arc   (d) axis

__D__  3. If no modifier is specified in the feature control frame, how is perpendicularity applied?
 (a) at LMC   (b) at MMC
 (c) at virtual condition   (d) RFS

__D__  4. What type of feature can have MMC applied with perpendicularity?
 (a) a tab   (b) a slot   (c) a keyway   (d) all of the above

__D__  5. What is the allowance based on, for two mating parts, when they are controlled with perpendicularity?
 (a) MMC   (b) LMC   (c) RFS   (d) virtual condition

__D__  6. What additional characteristic is controlled when perpendicularity is applied to a plane surface?
 (a) surface finish   (b) circularity
 (c) parallelism   (d) flatness

# Chapter 13 Review Problems, Part II

1. A surface of a part is perpendicular to another surface within .003. In the space below, draw the appropriate feature control frame.

   | ⊥ | .003 | A |

2. A hole in a part is perpendicular with the flat base within .002, regardless of the produced size of the hole. In the space below, draw the appropriate feature control frame.

   | ⊥ | ⌀ .002 | A |

3. In the part shown below, the hole at its MMC is perpendicular with the cylinder, RFS, within .004. Add an appropriate dimension and feature control frame.

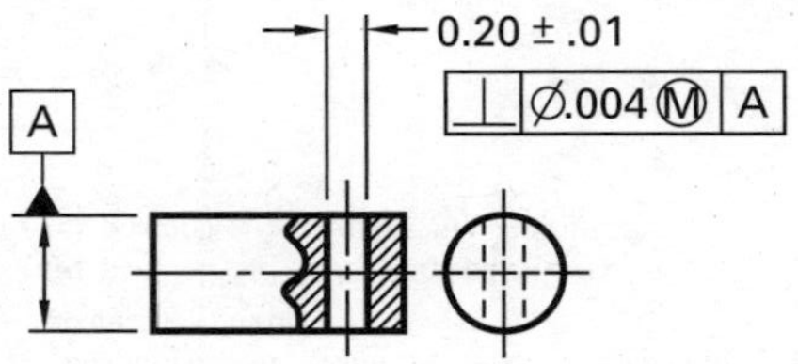

   — 0.20 ± .01

   | ⊥ | ⌀ .004 Ⓜ | A |

4. In problem 3, assume that the hole is dimensioned .250–.255 diameter. The perpendicularity tolerance will vary with the produced size of the hole. Complete the tolerance variation table below for every one-thousandth variation of the hole size.

   | Produced Size of Hole | Perpendicularity Tolerance |
   |---|---|
   | .250 | ⌀ .002 |
   | .251 | .003 |
   | .252 | .004 |
   | .253 | .005 |
   | .254 | .006 |
   | .255 | .007 |

5. In problem 4, what effect does the produced size of the cylinder have on the perpendicularity tolerance?

   _none_

# Chapter 14 Review Problems, Part I

Write the letter of the best answer in the space at left.

__A__  1. What is the shape of the tolerance zone when the tolerance is expressed in degrees?
 (a) fan-shaped  (b) uniform  (c) radial  (d) conical

__C__  2. What is the shape of an angularity tolerance zone?
 (a) angular  (b) fan-shaped  (c) uniform  (d) radial

__C__  3. Which of the following is not a possible form for angularity?
 (a) parallel lines  (b) parallel planes
 (c) parallel curves  (d) cylinder

__D__  4. What is the connection between datums and angularity tolerances?
 (a) may use  (b) should use
 (c) should not use  (d) must use

__A__  5. How is angularity dimension noted on the drawing for angularity?
 (a) basic  (b) limit form  (c) bilateral  (d) unilateral

__C__  6. How is angularity applied unless otherwise specified?
 (a) at LMC  (b) at MMC
 (c) RFS  (d) at virtual condition

__C__  7. Angularity of size features are more often controlled using which geometric control?
 (a) parallelism  (b) perpendicularity
 (c) position  (d) profile of a surface

6. In the drawing below, show that the boss is perpendicular to the base within .002 at MMC.

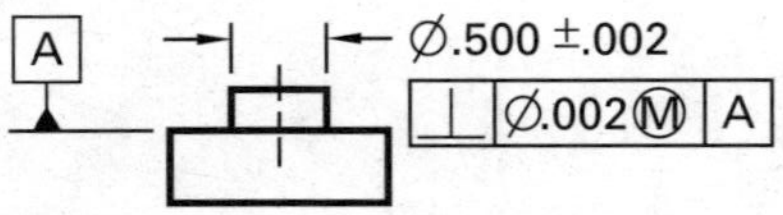

⌀.500 ±.002
⟂ | ⌀.002 Ⓜ | A

7. For the drawing in Problem 6, make up a tolerance variation table for five possible produced sizes of the boss. Use your own headings.

| Actual Size | Perpendicularity Tolerance |
|---|---|
| .502 | .002 |
| .501 | .003 |
| .500 | .004 |
| .499 | .005 |
| .498 | .006 |

8. The two drawings below show a part with a tab and a part with a mating slot.

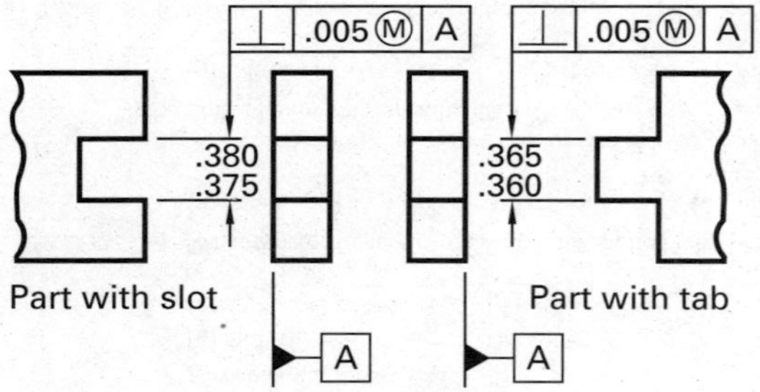

.380 / .375
⟂ | .005 Ⓜ | A

.365 / .360
⟂ | .005 Ⓜ | A

Part with slot         Part with tab

(a) What is the MMC of the slot?  .375
(b) What is the MMC of the tab?  .365
(c) What is the virtual size of the slot?  .370
(d) What is the virtual size of the tab?  .370
(e) What is the minimum clearance between the tab and the slot at the tightest fit?  0

VC TAB = MMC TAB + PER TOL
.370 = .365 + .005

.375 MMC SLOT
−.005 PER TOL
.370 VC

# Chapter 14 Review Problems, Part II

1. In the part shown below, the slanted surface is held at 30° to the base within .005. Add an angularity tolerance to complete the drawing.

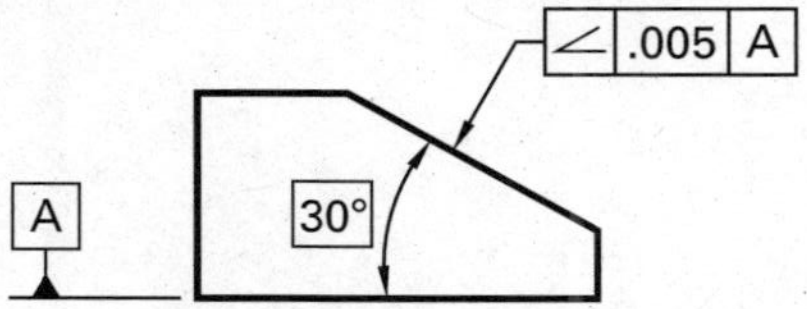

2. In the part shown below, the hole, regardless of feature size, is held at 60° to the base within .005. Add an angularity tolerance to complete the drawing.

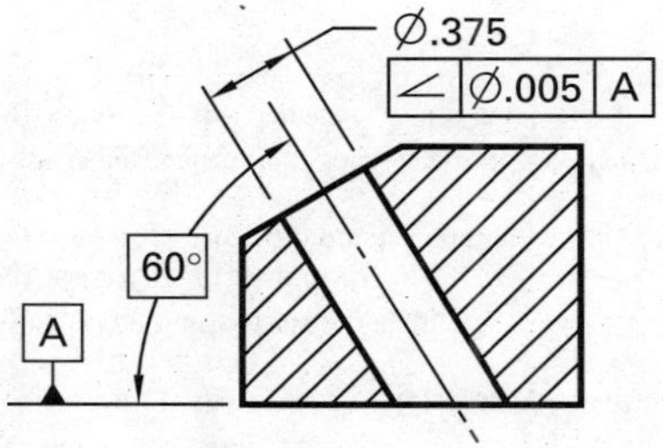

3. In the space below, draw the correct feature control frame for the drawing in problem 2 if the angularity tolerance were held at the MMC of the hole.

4. A .375–.380 diameter hole has an angularity tolerance of .006 at MMC relative to another hole, RFS. In the space below, draw a feature control frame to specify this situation.

5. When the .375–.380 diameter hole in problem 4 is *not* produced at its MMC, the angularity tolerance will not be .006. Make up a tolerance variation table below, giving the permissible angularity for each produced hole size, in thousandths.

| Produced Size of Hole | Angularity Tolerance |
|---|---|
| .375 | .006 |
| .376 | .007 |
| .377 | .008 |
| .378 | .009 |
| .379 | .010 |
| .380 | .011 |

6. In the drawing in problem 1, the flatness of the slanted surface is not specified. What is the maximum permissible out-of-flatness of this surface?

.005

# Chapter 15 Review Problems, Part I

Write the letter of the best answer in the space at left.

**C** 1. What word is generally used to describe two circular features in line with each other?
(a) concentricity  (b) alignment
(c) coaxiality  (d) runout

**B** 2. In addition to concentricity and runout, which geometric characteristic can control the axial alignment of a hole?
(a) location  (b) position
(c) cylindricity  (d) circularity

**A** 3. Concentricity error is the amount by which the axes of two regular solids are _____ .
(a) out of line  (b) in alignment  (c) curved  (d) irregular

**D** 4. Which of the three geometric characteristics used to control coaxiality is used the least?
(a) position  (b) runout
(c) circularity  (d) concentricity

**A** 5. The form of the concentricity tolerance zone is an imaginary cylinder about the axis of the _____ .
(a) datum  (b) center  (c) feature  (d) outside diameter

**C** 6. How is concentricity always applied?
(a) at LMC  (b) at MMC
(c) RFS  (d) at virtual condition

**B** 7. How is the concentricity tolerance zone specified?
(a) radius  (b) diameter  (c) total width  (d) radial width

# Chapter 15 Review Problems, Part II

1. In the part below, the smaller diameter is to be concentric with the larger diameter within .002. Complete the drawing.

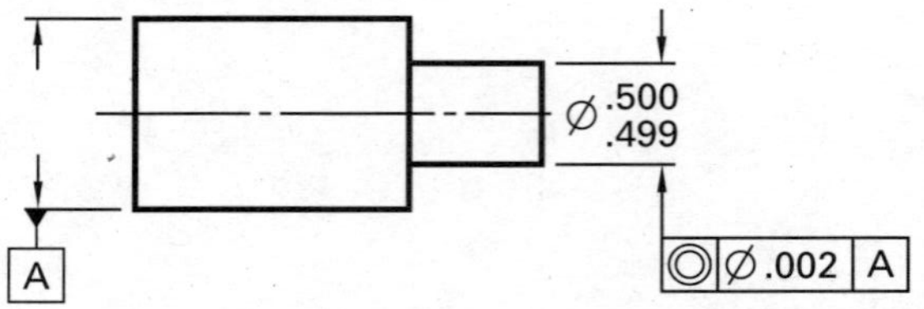

2. In the part below, the ⌀.998–1.000 is to be concentric within .001 with the two small cylinders, used as one datum. Complete the drawing.

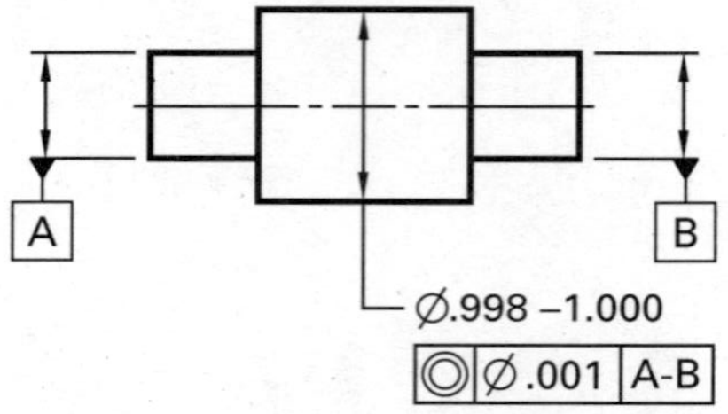

3. The drawing below is an incomplete interpretation of the concentricity control in the part shown in problem 2. Add an appropriate local note at each leader to complete the interpretation.

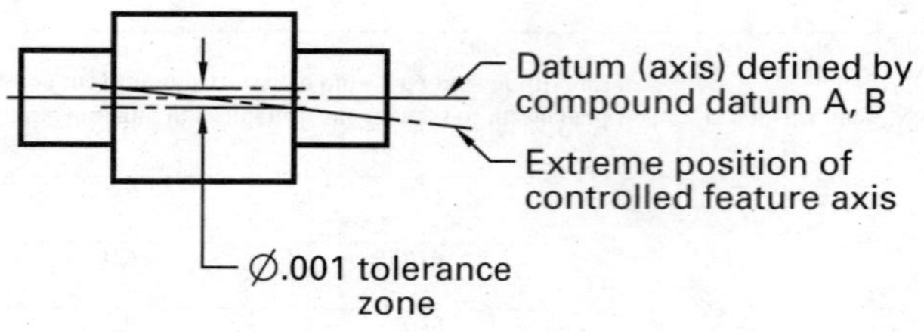

78  Answers to Review Problems  © Glencoe/McGraw-Hill

# Chapter 16 Review Problems, Part I

Write the letter of the best answer in the space at left.

__B__ 1. How is runout inspected?
 (a) taking diametral measurements at 90°
 (b) rotating the part about an axis
 (c) using a micrometer
 (d) using a comparator

__A__ 2. What is the name for runout when it controls circularity, straightness, and coaxiality?
 (a) composite tolerance (b) complex tolerance
 (c) multiple tolerance (d) position tolerance

__D__ 3. Circular runout is a control of circular _____.
 (a) features (b) axes (c) surfaces (d) elements

__A__ 4. Total runout is a control of all surface _____.
 (a) elements (b) features (c) datums (d) irregularities

__B__ 5. Circular runout is more commonly used because it is adequate for most designs and it is less expensive to _____.
 (a) control (b) measure (c) specify (d) analyze

__C__ 6. The form of the tolerance zone for circular runout is the radial distance between theoretical _____.
 (a) cylinders (b) elements (c) circles (d) radii

__A__ 7. The form of the tolerance zone for total runout is the radial distance between theoretical _____.
 (a) cylinders (b) elements (c) circles (d) radii

__B__ 8. What type of feature can be selected as a primary runout datum when the secondary datum is a relatively large cylindrical diameter?
 (a) axis (b) flat face
 (c) pair of lathe centers (d) element

__C__ 9. How is runout always applied?
 (a) at LMC (b) at MMC
 (c) RFS (d) at virtual condition

# Chapter 16 Review Problems, Part II

1. The circular runout of a diameter is to be held to .002 relative to one datum axis established by two diameters, A and B. Draw the appropriate feature control frame.

 ↗ | .002 | A-B

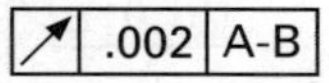

2. The total runout of a diameter has to be maintained within .003 relative to another diameter and a perpendicular face. Draw an appropriate feature control frame. (Perpendicularity will not be specified.)

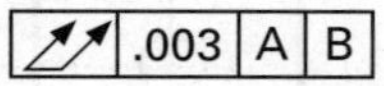

 ↗↗ | .003 | A | B

3. In the stepped shaft shown below, the circular runout of the conical surface relative to the smaller cylinder must be held within .001. Complete the drawing.

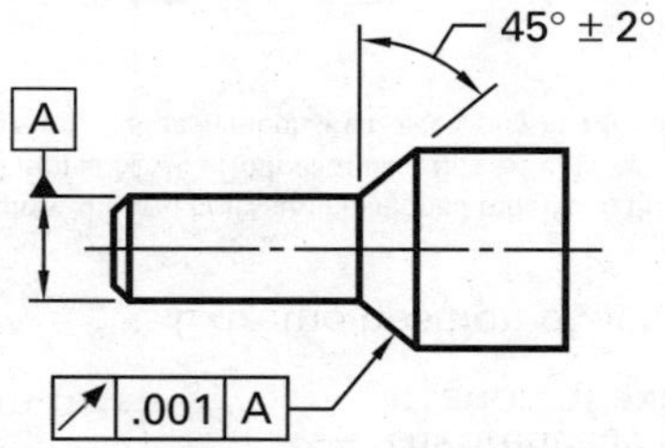

 45° ± 2°

 ↗ | .001 | A

4. Under what two conditions is it best to use runout to control coaxiality rather than concentricity or position tolerance?
 1. *Surface errors may be included.*
 2. *RFS applies to datum feature and controlled feature.*

## Chapter 17 Review Problems, Part I

Write the letter of the best answer in the space at left.

__A__  1. What is the relationship between a feature axis or center plane relative to other features for position?
   (a) theoretically exact   (b) approximate
   (c) average   (d) accurate

__A__  2. What type of error does position allow?
   (a) permissible   (b) measurable
   (c) unilateral   (d) bilateral

__A__  3. What is the name for position when it controls location, parallelism, and perpendicularity at the same time?
   (a) composite tolerance   (b) complex tolerance
   (c) multiple tolerance   (d) profile tolerance

__D__  4. How is position usually applied when controlling circular features?
   (a) half-width   (b) total width   (c) radius   (d) diameter

__A__  5. When position is applied to rectangular features, what must remain within the tolerance zone?
   (a) center plane   (b) center line of symmetry
   (c) axis   (d) boundary

__C__  6. What would be an equivalent coordinate square tolerance zone for a .014 diameter position tolerance zone that would not allow interference between the mating parts?
   (a) .006   (b) .007   (c) .010   (d) .014

__B__  7. What type of dimensions are used with position?
   (a) chain   (b) basic   (c) limit   (d) toleranced

__C__  8. If a hole is positioned within a .014 diameter tolerance zone, how much can it be out-of-perpendicular?
   (a) .0035   (b) .007   (c) .014   (d) .021

## Chapter 17 Review Problems, Part II

1. A part contains a number of holes that are positioned relative to two edges of the part within ⌀0.12 mm at MMC. Draw an appropriate feature control frame.

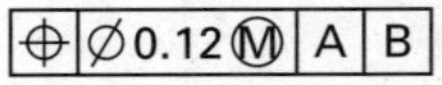

⊕ | ⌀0.12 Ⓜ | A | B

2. The drawing below is the interpretation for a part with a large hole held in position relative to two edges. Complete the local notes.

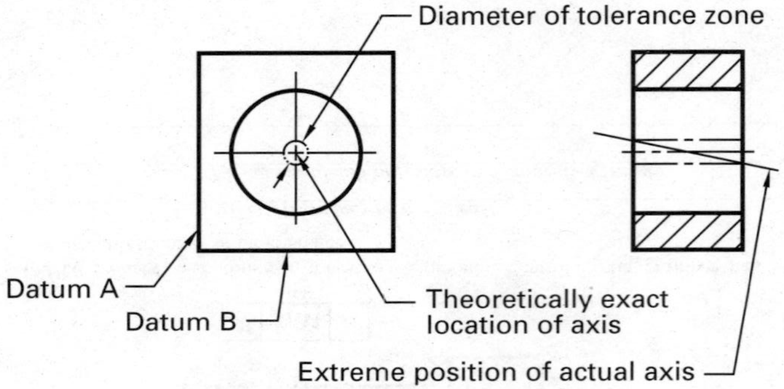

- Diameter of tolerance zone
- Theoretically exact location of axis
- Extreme position of actual axis
- Datum A
- Datum B

3. In the drawing below, the two holes are positioned relative to the left and bottom edges within ⌀0.3 mm at MMC. Both datum surfaces are flat within .15 mm. Complete the drawing, adding symbolism as necessary and location dimensions for the holes.

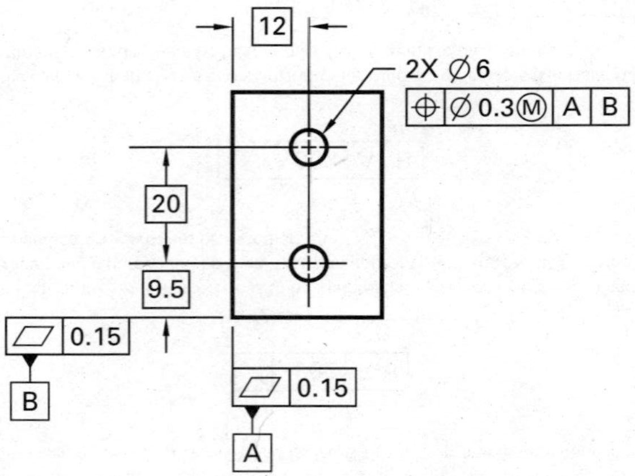

2X ⌀6
⊕ | ⌀0.3 Ⓜ | A | B

4. A strap has four 5 mm ± .05 mm punched holes in line, with a position tolerance of ⌀.3 mm at MMC. Complete the tolerance variation table below for the possible produced sizes given.

| Produced Size of Hole | Position Error Permissible |
|---|---|
| ⌀5.05 mm | 0.40 |
| 5.02 | 0.37 |
| 5.00 | 0.35 |
| 4.98 | 0.33 |
| 4.97 | 0.32 |

MMC = 4.95    0.02 bonus
LMC = 5.05    + 0.3  GT
              0.32 final tol

4.97 actual size
−4.95 MMC
 0.02 bonus

5. Relative to the example below, describe what happens to the four holes if the large hole is not at MMC.

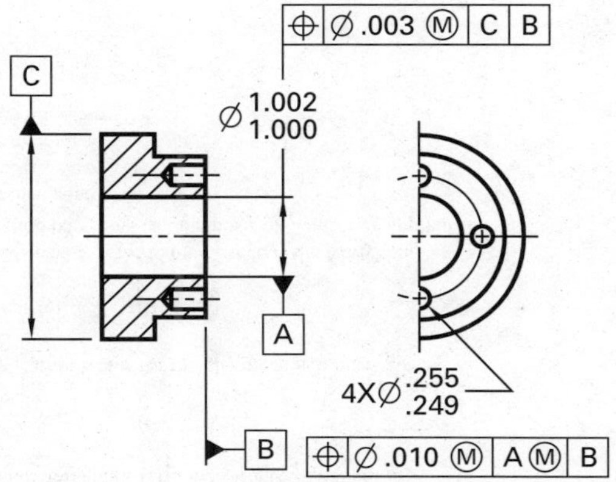

As datum A increases in size, the clearance between the mating feature and datum A also increases. The four holes are controlled relative to datum A at MMC. Therefore, if there is an increase in clearance, the four holes together (as a group) will also experience this additional clearance, as a shift from the theoretical center of datum A.

The drawing of the spacer below is an application of the control of an edge distance by specifying LMC.

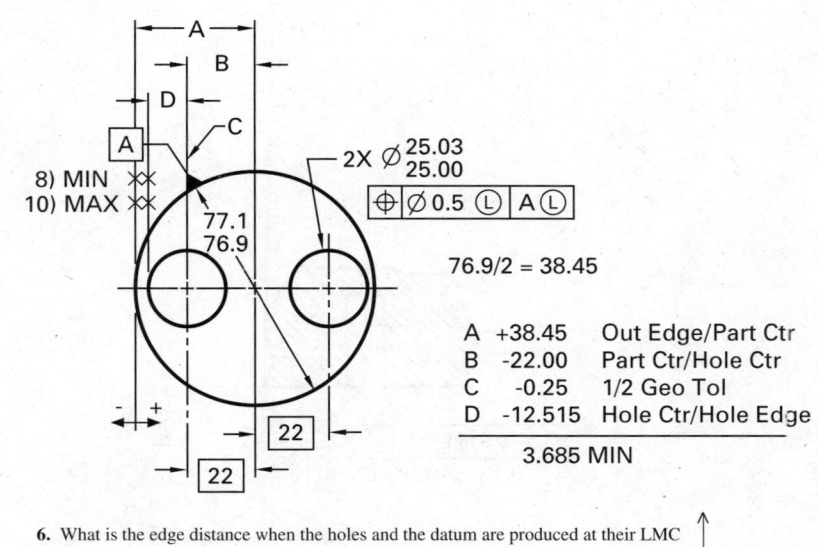

8) MIN
10) MAX

A  +38.45   Out Edge/Part Ctr
B  −22.00   Part Ctr/Hole Ctr
C   −0.25   1/2 Geo Tol
D  −12.515  Hole Ctr/Hole Edge
    3.685 MIN

76.9/2 = 38.45

6. What is the edge distance when the holes and the datum are produced at their LMC sizes?

3.685

7. What is the maximum position tolerance for the two holes when they and the datum are *not* at LMC?

  .05 GT
+ 0.03 bonus
+ 0.2  shift
  0.73 total

8. If the two holes and the datum were produced at their MMC sizes, what would be the edge distance?

77.1
−76.9
  0.2

77½ = 38.55

+ 38.55   OD edge/part ctr
−  22     Part ctr/hole ctr
−  0.25   ½ geo tol
−  0.015  ½ bonus
−  0.1    ½ shift
− 12.5    hole ctr/hole edge
  3.685

## Chapter 18 Review Problems, Part I

Write the letter of the best answer in the space at left.

_C_ 1. What type of holes do both parts have in a floating fastener assembly?
 (a) close-fitting  (b) press-fit  (c) clearance  (d) tapped

_A_ 2. In a fixed fastener assembly of two parts, one of the mating parts has a clearance hole. What type of hole does the other part have?
 (a) tight-fitting  (b) loose-fitting
 (c) transition fit  (d) lubricated

_A_ 3. Before a position tolerance can be specified, what must be calculated?
 (a) clearance  (b) perpendicularity
 (c) parallelism  (d) concentricity

_C_ 4. What is the multiplication factor for the clearance in a floating fastener calculation?
 (a) $\frac{1}{4}$  (b) $\frac{1}{2}$  (c) 1  (d) 2

_B_ 5. What is the multiplication factor for the clearance in a fixed fastener calculation?
 (a) $\frac{1}{4}$  (b) $\frac{1}{2}$  (c) 1  (d) 2

_B_ 6. If a part is rejected because a size feature exceeds the MMC but actually fits the mating part, how can the position control be applied?
 (a) projected tolerance zone
 (b) zero position tolerance at MMC
 (c) larger position tolerance
 (d) smaller position tolerance

## Chapter 18 Review Problems, Part II

1. Write a local note to replace the feature control frame in the drawing below.

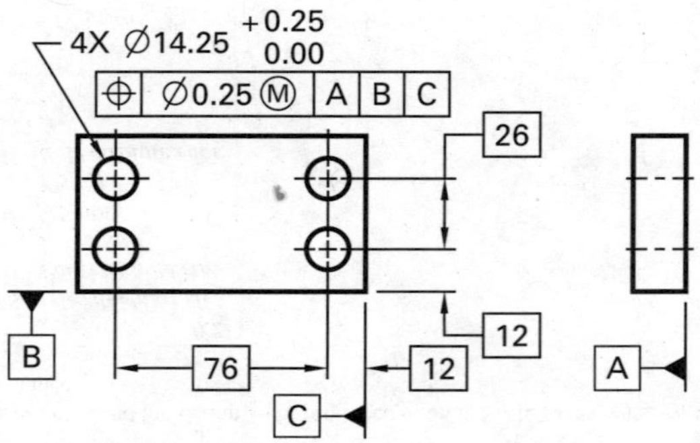

*The feature axes must be within a 0.25 diameter tolerance zone at MMC, relative to datum features A, B, and C.*

2. To the drawing below, add a requirement that the axis of the hole be perpendicular to the top surface within ⌀.3 mm at MMC and a requirement that the tolerance be projected 14 mm above the top surface.

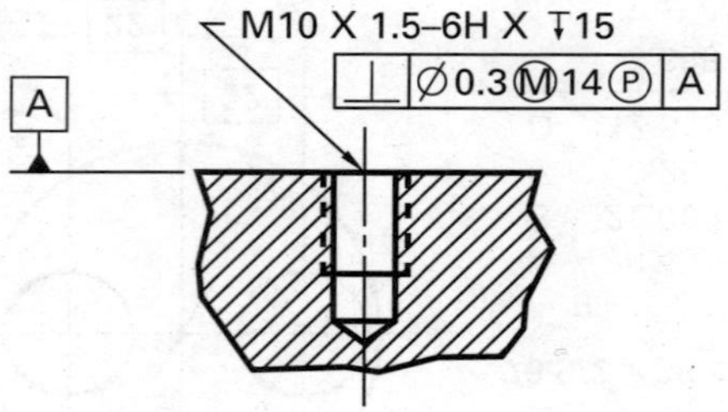

3. The next drawing is an application of zero position tolerance at MMC. The only variation allowed is in the size of the hole. Complete the tolerance variation table below for the given produced sizes of the hole.

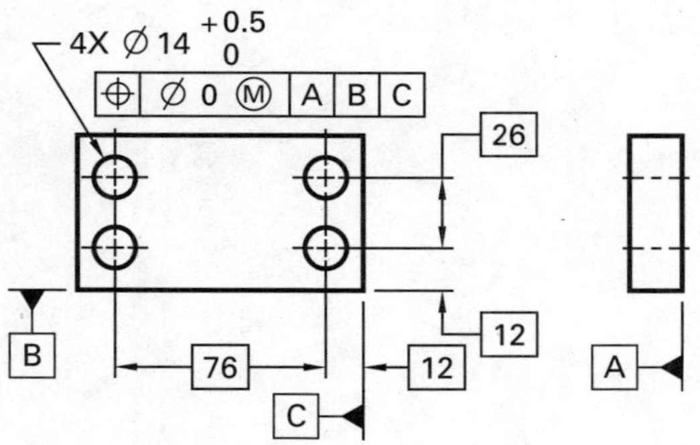

| Produced Hole Size (mm) | Permissible Position Tolerance (mm) |
|---|---|
| ⌀14.0 | 0 |
| ⌀14.2 | 0.2 |
| ⌀14.5 | 0.5 |

4. In the situation shown in problem 3, 13.9–14.0 mm screws are to be inserted in the four holes. What will be the allowance (tightest fit) between the screws and the holes when:

(a) the holes are produced at MMC?

```
 14.0 MMC hole
 14.0 MMC screw      14.5
    0               13.9
                     0.6
 14.4
-14.0
  0.4
```
   0

(b) the holes are produced at ⌀14.4 mm?   0.4

(c) What is the maximum possible clearance (loosest fit) between the screws and holes?   0.6

5. The formula used to determine the clearance and position tolerance in the design of mating tabs and slots is similar to the floating fastener formula. Write the formula below, using

$T_{tot}$ for total tolerance in both parts    $T_{tot} = W_s - W_t$

$W_s$ for width of slot at MMC

$W_t$ for width of tab at MMC

6. The drawing below shows a disk with two keyways located by position.

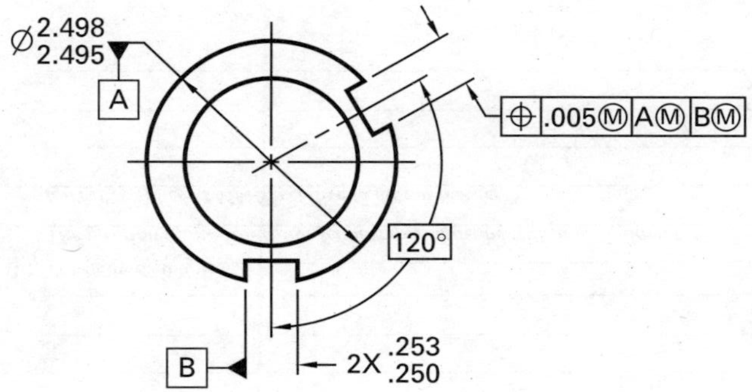

(a) What geometric characteristic (besides position) is actually controlled?   *angularity*

(b) Write a local note that could replace the feature control frame.
   *The position of the feature center plane must be within .005 at MMC relative to datum feature A at MMC and B at MMC.*

(c) Give one possible reason why the outside diameter is the more important of the two datums.
   *The OD may fit in a hole of the mating part.*

7. For the drawing in problem 6, complete the tolerance variation table below for the possible produced sizes shown. (Hint: The total departure from MMC for all the sizes involved must be found.)

| Produced Size | | | | Permissible | Stated | Total Depart |
| Upper Slot | Lower Slot | Outside Diameter (O.D.) | | Position Tolerance | Tol | from MMC |
|---|---|---|---|---|---|---|
| .253  .003 | .253  .003 | 2.498  0 | | .011 | .005 | + .006 |
| .252  .002 | .252  .002 | 2.497  .001 | | .010 | .005 | + .005 |
| .251  .001 | .251  .001 | 2.496  .002 | | .009 | .005 | + .004 |
| .250   0  | .250   0  | 2.495  .003 | | .008 | .005 | + .003 |

departure from MMC

9. Write notes to interpret the composite feature control frame for the circular hole pattern in the drawing below.

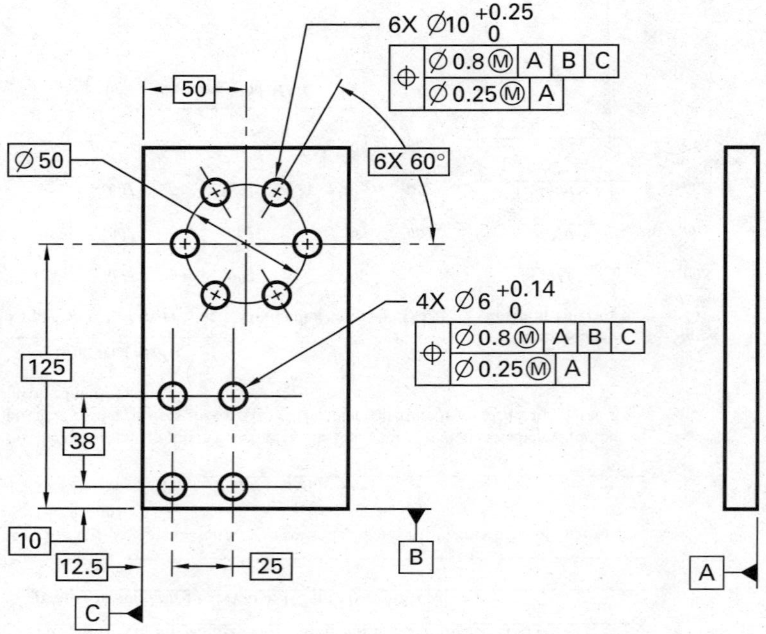

(a) The pattern location
*The position of the pattern of feature axes must be within 0.8 diameter tolerance zones at MMC, relative to datum features A, B, and C.*

(b) The individual holes
*The position of the feature-to-feature axes must be within 0.25 diameter tolerance zones at MMC, relative to datum feature A.*

8. The following questions refer to the drawing below of a plate with multiple patterns of features (two hole patterns).

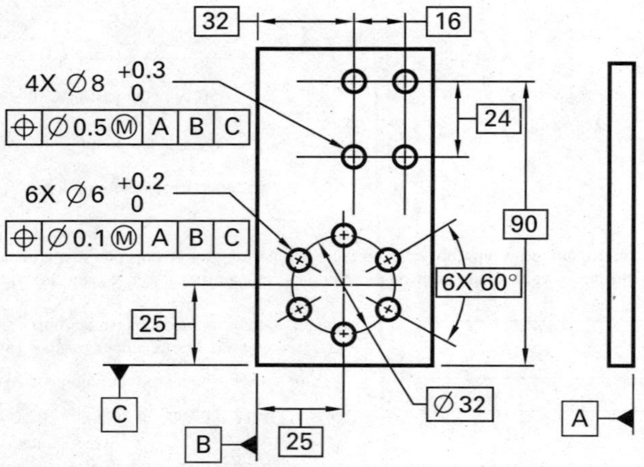

(a) Should the hole patterns be inspected as (check one) _____ separate patterns? __X__ one composite pattern?

(b) Give the rule concerning multiple patterns of features that supports your answer to problem 8 (a).
*When multiple patterns of features are located realtive to the same datums, they are considered one pattern.*

(c) How many gages will be used to inspect the location of the holes? __one__

# Chapter 19 Review Problems, Part I

Write the letter of the best answer in the space at left.

_A_ 1. What do two or more coaxial features share?
(a) center  (b) center plane  (c) axis  (d) datum

_B_ 2. What is the shape of the tolerance zone used for coaxial position?
(a) circle  (b) cylinder  (c) ring  (d) tube

_B_ 3. How is a coaxial position tolerance specified?
(a) radius  (b) diameter  (c) half-width  (d) total width

_B_ 4. Which modifier is used for coaxiality to ensure all parts will assemble?
(a) LMC  (b) MMC  (c) RFS  (d) virtual condition

_A_ 5. In addition to controlling wall thickness, what else can an LMC modifier control?
(a) edge distance  (b) basic dimension
(c) location dimension  (d) assemblability

_A_ 6. What is the total position tolerance equal to for a stepped shaft (two diameters) and stepped sleeve assembly?
(a) sum of the maximum clearances
(b) sum of the minimum clearances
(c) difference between the larger diameters
(d) difference between the smaller diameters

_A_ 7. What is avoided when using zero position tolerancing?
(a) unused position tolerance  (b) excess clearance
(c) insufficient clearance  (d) excess runout

# Chapter 19 Review Problems, Part II

1. In the seal body drawn below, the small diameter must be coaxial with the large diameter within ⌀.001 when both diameters are at MMC. Add a position tolerance to complete the drawing.

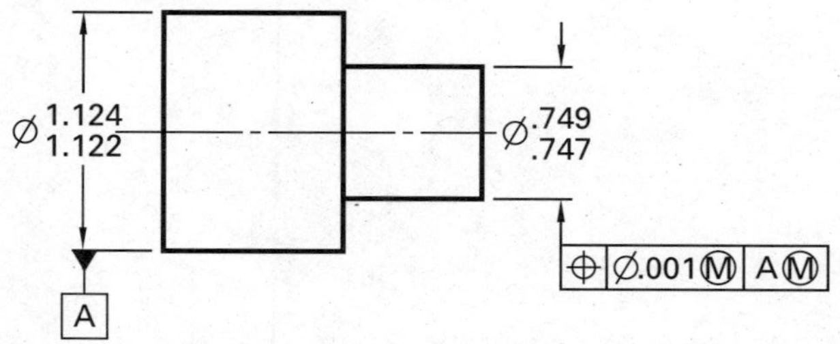

2. The outside diameter of a thin-wall sleeve must be positioned within ⌀.1 mm at LMC relative to the inside diameter at LMC. Add a position tolerance to complete the drawing.

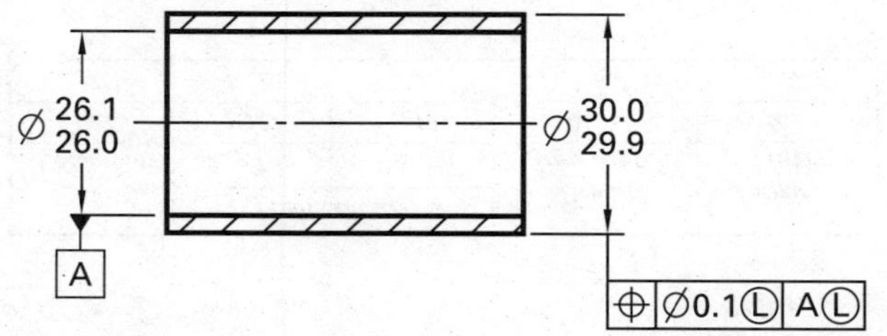

3. The small diameter of the balance block shown below must be coaxial within ⌀.0005 at MMC with the large diameter, no matter what the produced size of the large diameter might be. Also, the large diameter at MMC must be positioned within ⌀.010 relative to the height and width of the block at their MMC sizes. Add a position tolerance to complete the drawing.

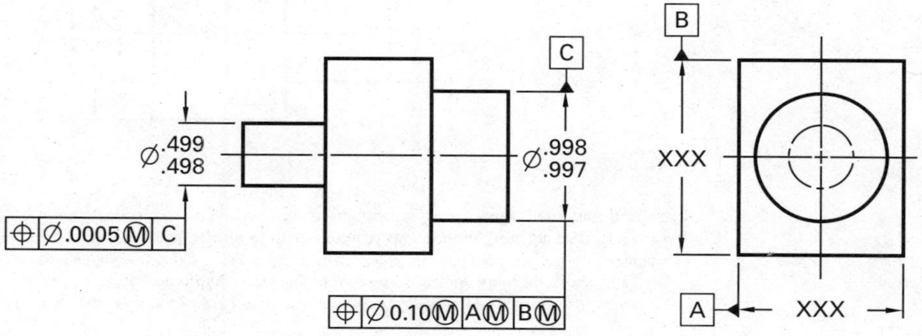

4. In the hinge body below, the four ⌀10-mm holes are to be coaxial within ⌀.1 mm at MMC and positioned within ⌀.2 mm at MMC relative to two perpendicular flat surfaces. Add a composite feature control frame and datum feature symbols to express this design intent.

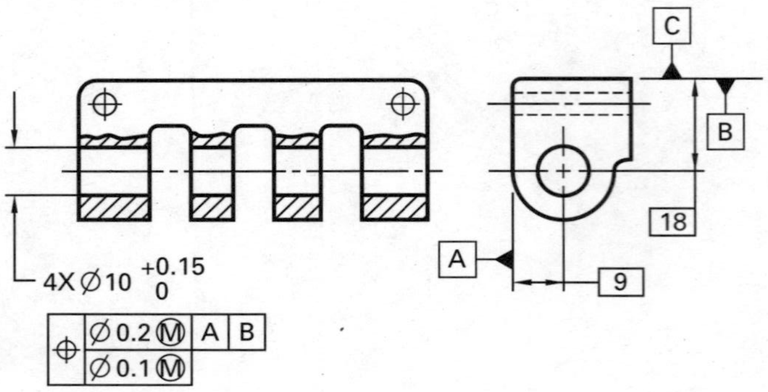

Tolerance on plug: .001

Do not forget to add the position tolerances to the drawings.

6. The same seal body as in problem 1 is drawn below but with a zero position tolerance at MMC controlling coaxiality of the two diameters. Make up a tolerance variation table for two possible produced sizes: the MMC and the LMC for both diameters. Show the total departure (if any) from MMC resulting from the produced sizes. Show also the resulting permissible position error. Neatly letter appropriate headings.

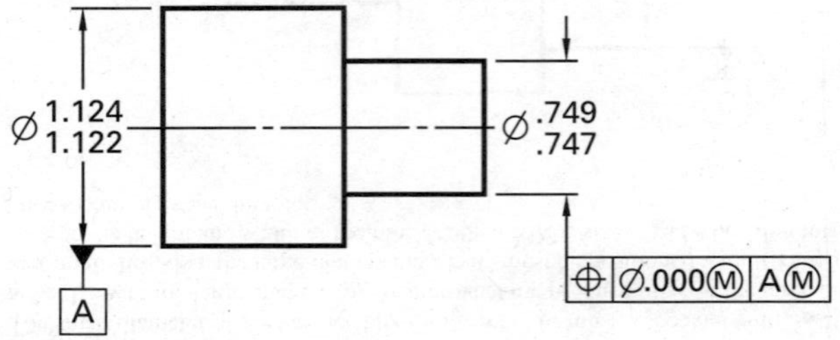

| Condition | Produced Size | | Depart from MMC | Permissible Error |
|---|---|---|---|---|
| | Lrg Dia | Sm Dia | | |
| MMC | 1.124 | .749 | 0 | 0 |
| LMC | 1.122 | .747 | .004 | .004 |

5. Below are two drawings, one each of a socket and a mating plug. Use the three-step procedure to calculate the position tolerance for each part, making the tolerance for the socket larger than the plug tolerance. Add the position tolerances to the drawings.

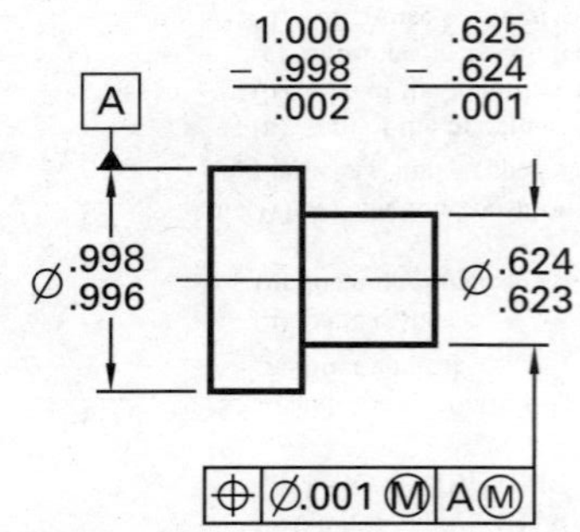

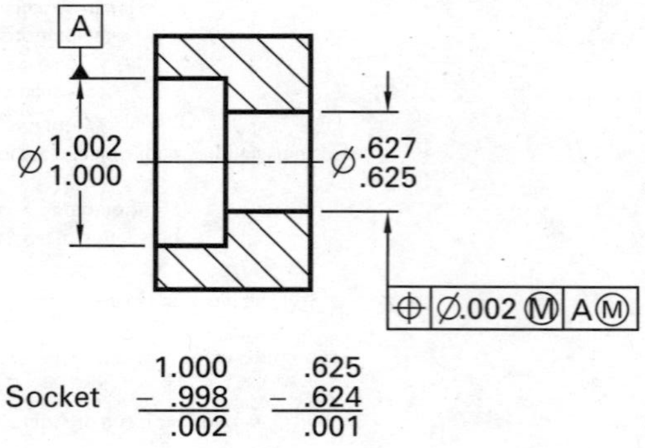

Step 1: Obtain the minimum clearances

Small diameters: .001      Large diameters: .002

Step 2: Add the minimum clearances to obtain the total position tolerance for both parts.

The sum is .003

Step 3: Divide the total position tolerance between the two parts.

Tolerance on socket: .002

Plug

86    Answers to Review Problems                              © Glencoe/McGraw-Hill

# Chapter 20 Review Problems

Write the letter of the best answer in the space at left.

__D__  1. Symmetry is used in which of the following conditions?

    (a) mating parts    (b) interchangeable fits
    (c) force fits    (d) individual part

__C__  2. What is the main concern for the designer when using symmetry?

    (a) wall thickness    (b) strength
    (c) balance    (d) all of the above

__B__  3. Why is symmetry not commonly used?

    (a) it applies only to force fits    (b) it is expensive to inspect
    (c) it is hard to understand    (d) there are no tolerance calculations

# Answers to Comprehensive Exercises

| Exercise | Answer Page |
|---|---|
| 1 | 91 |
| 2 | 92 |
| 3 | 93 |
| 4 | 93 |
| 5 | 94 |
| 6 | 94 |

# Comprehensive Exercise 1

## Definitions and Symbols

In the parentheses in the right-hand column, place the number of the matching term from the left-hand column. Use every term in the left-hand column.

1. Angularity
2. Basic dimension
3. Bilateral tolerance
4. Concentricity
5. Cylindricity
6. Datum
7. Flatness
8. Limits
9. Maximum material condition
10. Parallelism
11. Perpendicularity
12. Position
13. Profile of a line
14. Profile of a surface
15. Regardless of feature size
16. Circularity
17. Runout, circular
18. Runout, total
19. Straightness
20. Tolerance
21. Unilateral tolerance
22. Symmetry

(15) Where the tolerance of form or position must be met, regardless of where the feature lies within its size tolerance.
(10) ∥
(5) ⌭
(2) The theoretical value used to describe the exact size, shape, or location of a feature.
(7) ▱
(1) ∠
(3) One in which variation is permitted in both directions from the specified dimension.
(21) One in which variation is permitted in only one direction from the specified dimension.
(14) ⌒
(18) ⤰
(4) ◎
(17) ↗
(12) ⊕
(9) The condition of a part feature when it contains the maximum amount of material.
(11) ⊥
(19) —
(20) The total amount by which a dimension may vary.
(16) ○
(13) ⌒
(8) The maximum and minimum sizes indicated by a toleranced dimension.
(6) A surface indicated on the drawing from which measurements can be made.
(22) ≡

# Comprehensive Exercise 2

## Applications of Geometric Tolerances

Add feature control frames and datum identifying symbols to the drawing below to specify the following requirements. All symbols and feature control frames should be sized correctly.

1. Surface 1 to be circular within .005.
2. Surface 4 to be flat within .001.
3. Circular runout of surface 2 within .01 relative to surfaces 4 and 1.
4. Total runout of surface 6 within .003 relative to surfaces 4 and 1.
5. Surface 5 to be parallel within .001 to surface 4.
6. Position the 1.50-12 threads RFS within .005 to surface 4 and surface 1 at MMC.
7. Surface 3 to be positioned within .002 diameter at MMC relative to the 1.50-12 threads RFS.

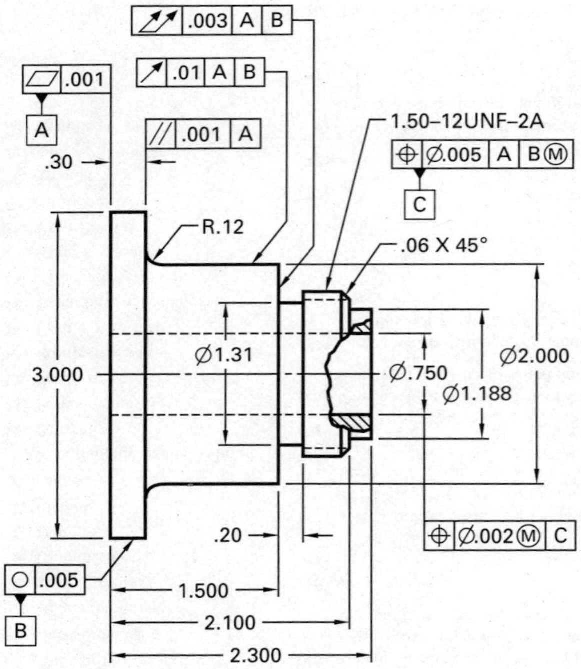

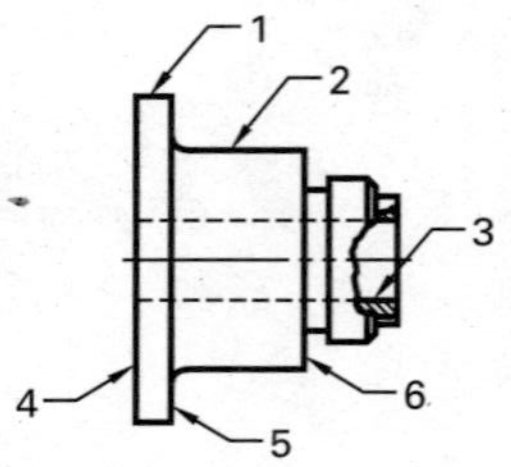

For each feature control frame and datum symbol on the previous page, write a note that might be used in place of the symbols. List them below by the numbers corresponding to the numbers shown on the drawing.

1. The circularity of the feature must be within .005.
2. The flatness of the feature must be within .001.
3. The circular runout of the feature must be within .01 relative to datums A and B.
4. The total runout of the feature must be within .003 relative to datums A and B.
5. The parallelism of the feature must be within .001 relative to datum A.
6. The position of the feature axis must be within a cylindrical tolerance zone of .005 diameter RFS relative to datum A and datum B at MMC.
7. The position of the feature axis must be within a cylindrical tolerance zone of .002 diameter at MMC relative to datum C RFS.

# Comprehensive Exercise 3

## Calculations for Fasteners

**1.** A lid assembles onto a gear case with six $\frac{5}{16}$–18 cap screws. Using a position tolerance of ⌀.015 at MMC for the screw hole locations in both parts, calculate the size of the clearance holes in the lid. Select a drill size from the standard sizes listed at right. There are internal threads in the gear case. Perform all calculations neatly, starting with a formula.

*Fixed fastener*

$GT = \frac{H-F}{2}$

$.015 = \frac{x - .3125}{2}$

$x = .3125 + (2 \times .015)$

$x = .3425$

| Drill Sizes |
|---|
| .312 |
| .316 |
| .323 |
| .328 |
| .332 |
| .339 |
| .344 (circled) |
| .348 |
| .358 |

**2.** Three parts are held together with $\frac{1}{2}$–12 bolts. The clearance holes in all three parts are ⌀.531. What position tolerance at MMC will be required in the drawings of the three parts? Perform all calculations neatly, starting with a formula.

$GT = H - F$

$= .531 - .50$

$= .031$

# Comprehensive Exercise 4

## Calculations for Slots and Tabs

All the questions in this exercise pertain to the two drawings of mating parts below.

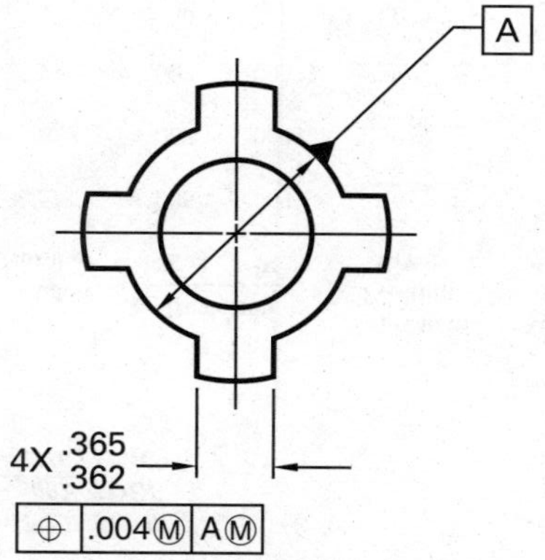

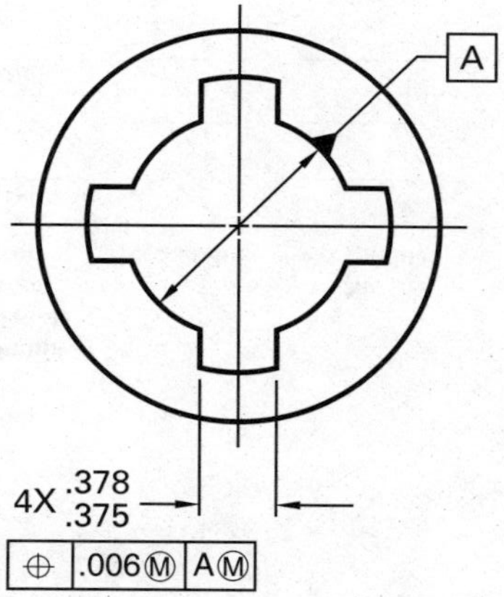

**1.** Calculate the position tolerance at MMC for the tabs and the slots, and add the appropriate feature control frames. Make the tab tolerance two-thirds of the slot tolerance. Show all calculations, starting with a formula.

Tab tolerance: __.004__    Slot tolerance: __.006__

$T_{TOT} = W_S - W_T$
$= .375 - .365$
$= .010$

.365 MMC
− .362 LMC
  .003 bonus
+ .004 stated GT
  .007

**2.** What is the maximum permissible position error when the tabs are at their LMC sizes? Show all calculations.

__.007__

**3.** What would be the maximum permissible error when the tabs are at their LMC sizes if the position tolerance were specified RFS? Show all calculations.

__.004__    *.004 tolerance applies, regardless of the size*

# Comprehensive Exercise 5

## Calculations for Fit of Coaxial Parts

You have been asked to design the fit of the socket and mating plug shown in the two drawings below.

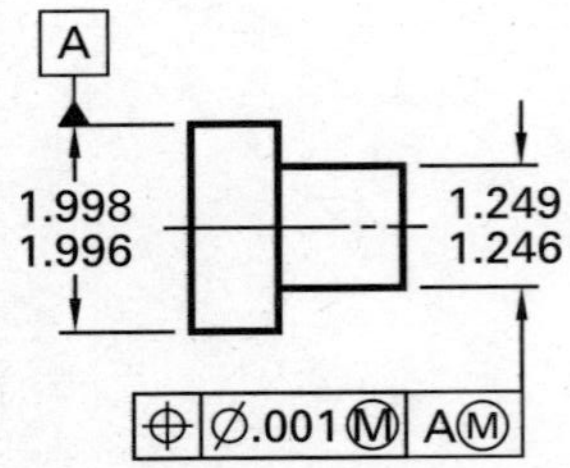

Plug—Scale: 1/2

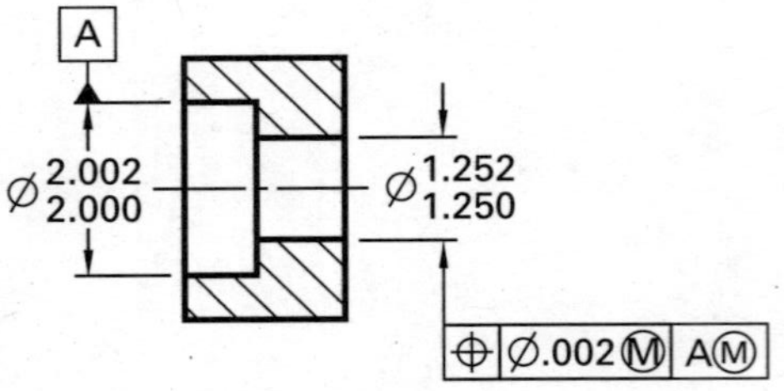

Socket—Scale: 1/2

Given:
The internal diameters of the socket:
   See drawing.
The tightest fit with permissible out-of-coaxiality:
   Line-to-line
The loosest fit with perfect coaxiality:
   .006
The coaxiality tolerances:
   Plug    ⌀.001
   Socket  ⌀.002

*Large Diameter*

| 2.000 | MIN | SOC |
|---|---|---|
| −.002 | MIN | CLEAR |
| 1.998 | MMC | PLUG |

| 2.002 | MAX | SOC |
|---|---|---|
| −.006 | MAX | CLEAR |
| 1.996 | LMC | PLUG |

*Small Diameter*

| 1.250 | | |
|---|---|---|
| −.001 | | |
| 1.249 | | |

| 1.252 | | |
|---|---|---|
| −.006 | | |
| 1.246 | | |

Required:
Decide whether the coaxiality tolerances and the datums should apply at MMC or RFS.
Calculate limit dimensions for the plug diameters. (There is more than one correct answer.) Calculations can be done on scratch paper.
Add the missing information to both drawings.
You are not concerned with dimensions not related to the fit.

# Comprehensive Exercise 6

## Conversion—Coordinate Tolerances to Position Tolerances

The answers to all the problems in this exercise can be found by using the conversion tables in Appendix B. Round off to the nearest thousandth unless otherwise specified.

In the inspection of a baseplate, a coordinate measuring machine was used to measure deviation from basic hole location dimensions from two perpendicular datums (X and Y), and the data given below were recorded. Convert each pair of values to equivalent round position tolerance values.

| Hole Number | Deviation X | Deviation Y | | Position Tolerance |
|---|---|---|---|---|
| 1. | .002 | .004 | ⇒ | .0089    .008 ✓ Always round down for geometric tolerances. |
| 2. | .008 | .006 | ⇒ | .0200    .020 |
| 3. | .005 | .001 | ⇒ | .0102    .010 |
| 4. | .007 | .003 | ⇒ | .0152    .015 |

A large number of 5/16-diameter holes are drilled in a housing. The print shows that the drill tolerance is +.006 −.002 and the holes must be positioned at MMC within .015 diameter. The deviations of the holes from their basic location dimensions are measured and recorded. The actual hole diameters are also measured and recorded. The table below presents the recorded data and provides spaces for other derived data to be filled in by the inspector. For each hole, obtain the equivalent position error at MMC, the *maximum permissible* position error, and indicate by a check mark whether the hole location is acceptable or not.

   .3125 BASIC
  − .002  TOL
   .3105 MMC

| Hole Number | Deviation X | Deviation Y | Equivalent Position Error | Actual Hole Size | Maximum Permissible Position Error | Acceptable | Not Acceptable |
|---|---|---|---|---|---|---|---|
| 5. | .006 | .005 | .015 | .310 | .015 | ✓ | |
| 6. | .007 | .004 | .016 | .312 | .017 | ✓ | |
| 7. | .002 | .003 | .007 | .315 | .020 | ✓ | |
| 8. | .008 | .006 | .020 | .318 | .023 | ✓ | |
| 9. | .007 | .008 | .021 | .314 | .019 | | ✓ |

94    Answers to Comprehensive Exercises     © Glencoe/McGraw-Hill